STUDY GUIDE/WORKBOOK

to accompany

SPEECH AND HEARING SCIENCE ANATOMY AND PHYSIOLOGY

FOURTH EDITION, W. R. ZEMLIN

Corrected Printing

EILEEN ZEMLIN, M.ED.
CCC S/LP

W. R. ZEMLIN
Professor Emeritus of Speech and Hearing Science
and School of Basic Medical Sciences
University of Illinois

Illustrations by

BARBARA DILLABAUGH SALINGER
THERESE ZEMLIN

Published by
STIPES PUBLISHING L.L.C.
204 W. University Ave.
Champaign IL 61820

To future professionals in
speech, language, and hearing

Table of Contents

Chapter 6: Hearing 231

Chapter 7: Embryology of the Speech and Hearing Mechanism 278

Preface

This study guide/workbook will be used primarily in courses for future professionals in speech, language, and hearing. It will be to your advantage to have had introductory courses in anatomy, physiology, and the physical sciences, but the text and study guide do not assume you have a scientific background. It will also be to your advantage to have anatomical specimens, anatomical models, and laboratory equipment available for demonstration and "hands-on" experience. Again, the text and study guide do not assume you have ready access to anatomy laboratories or fully equipped speech and hearing science laboratories.

As stated in the text, your primary goal in this course should be to develop "a *hypothetical construct* of the structure and function of the speech and hearing mechanism." A construct may be defined as a complex image or idea resulting from a synthesis by the mind. To determine whether or not you are establishing "realistically and clinically useful hypothetical and working constructs" you may ask yourself the following questions:

1. Can I visualize the anatomical structures, their growth, and the effects of aging?
2. Can I visualize and understand how these structures function under usual conditions, and hypothesize how they might function under adverse conditions?
3. Can I readily explain what I have learned to someone else?
4. Can I integrate various constructs into a meaningful, all-encompassing construct?
5. Will I be able to modify my construct when new research findings are presented?

We have attempted to provide comprehensive and relatively detailed information upon which to base your constructs of structures and functions, but once they have been established you needn't be concerned about retaining the details that helped you develop them.

The study guide/workbook, used in conjunction with the text, also attempts to:

1. encourage the development of a working vocabulary and an appreciation of the logic of that vocabulary.
2. demonstrate the clinical relevance of specific areas of study.
3. reveal the variability and complexity of the speech and hearing mechanisms and develop an appreciation of how much there is yet to learn about the structures and their functions.
4. provide some of the tools that will enable you to profit from reading professional journals and other literature in your field.

Because the last two chapters in the text are primarily for reference, only a few exercises on embryology have been included in chapter seven. We have intentionally provided relatively broad coverage of most areas, assuming instructors will recommend you emphasize some topics and de-emphasize others.

We would like to acknowledge Stafford Thomas and Jacqueline Schaffer for the use of their artwork, and Elaine Paden for giving us the benefit of her professional expertise. We would also like to thank Nancy Wallace and Lou Ann Koebel for their many contributions to the production of this edition.

Each author we have quoted has unknowingly made a very special contribution: clarifying a point, providing a graphic description, giving well-founded advice, sharing research findings, stimulating interest. To these authors we owe a particular debt of gratitude.

Eileen and W. R. Zemlin

Instructions for Use of the Study Guide

The exercises will be more useful if you write the short answers in the column provided at the side of the page. Then as you review, you will be able to check the answers as you slide a card down the column. Because the format of the questions is relatively consistent throughout the book and because several different types of questions may be used within one exercise, specific directions are not included in each exercise.

Direct Questions

1. Which months have only thirty days?

   Note that when short answers are required, the number of blanks in the column usually indicates the number of answers needed.

   1. *April*
      June
      November
      September

2. Why is it warmer in the summer? *because of inclination of earth rotating on its axis as it moves around the sun, sun's rays are less slanted in the summer.*

   Note that when longer answers are required, the number of lines does not indicate that more than one explanation is needed.

True-False

3. There are eight days in a week.

   3. T (F)

Fill in the Blanks

4. There are _____ days in a week and _____ weeks in a year.

   4. *7*
      52

5. The largest city in the United States is _____.

   5. *New York*

   Note that only one blank is used for a two-word answer.

Select the Correct Answer

6. Florida is in *the United States / Canada.*

   6. *U.S.*

7. New York is a *city / state / both.*

   7. *both*

   Note that both is used in lieu of city and a state. Proper grammatical usage has been sacrificed for the sake of brevity.

Matching

> bird
> cat
> dog

9. an animal that barks

9. _dog_

10. an animal that has fur

10. _dog_

    cat

Note that the two blanks in the answer column indicate the need for two answers.

11. an animal that flies

11. _bird_

12. an animal that has feathers

12. _bird_

The word lists in the matching exercises have been alphabetized, and every word in the list will be used <u>at least</u> once.

See **Answer Key** at the back of the book. Although the thought **Questions** are intended to stimulate thinking, you will find answers to a few of them at the very end of the answer section.

Suggestions:

Most of the illustrations are designed to serve as self tests. Although we hesitate to recommend coloring, it may be worthwhile if you concentrate not on beautifying the illustrations but on understanding what you are coloring.

We hope the self tests will (1) demonstrate how much you know or can readily surmise, (2) reveal how well you understand specific structures and their functions, (3) encourage your appreciation of the importance of organization and association, and (4) increase your active involvement in the learning process.

Goals:

When students in speech-language pathology begin their clinical practicum, they become exceedingly familiar with short-term and long-term goals. Comprehending any pertinent anatomical structure or physiological process will, in a sense, accomplish a short-term goal that contributes to your long-term goal of understanding the anatomical and physiological bases of speech, language and hearing. To maintain this understanding you must apply what you will learn in this course to your study of specific aspects or disorders of speech, language, and hearing. For a speech-language pathologist or audiologist, knowledge of basic structure and function is the solid core of professional competence.

Chapter 1
Introduction and Orientation

No. 1-1 GENERAL ANATOMICAL TERMS
Text pages 5–6

1. The bridge of the nose is *superior / inferior* to its tip.

   1. _superior_

2. In the anatomical position the thumb is *lateral / medial* to the little fingers.

   2. _lateral_

3. The shoulder is *distal / proximal* to the elbow which is *distal / proximal* to the wrist which is *distal / proximal* to the fingers.

   3. _proximal_

4. When the mouth is closed the lips are *anterior / posterior* to the front teeth which are *anterior / posterior* to the tip of the tongue.

   4. _anterior_
      anterior

5. Pain receptors in the skin are part of the *central / peripheral* nervous system.

   5. _central_

6. The spinal cord is part of the *central / peripheral* nervous system.

   6. _central_

7. A backpack rests on the *ventral / dorsal* surface of the body.

   7. _dorsal_

8. Muscle is *superficial / deep* in relation to skin.

   8. _deep_

9. The molars are *anterior / posterior* to the eyeteeth.

   9. _posterior_

10. The developing brain is located in the *rostral / caudal* region of the embryo.

    10. _rostral_

11. The umbilicus (belly button) is on the *ventral / dorsal* surface of the body.

    11. _ventral_

12. Beginning ice skaters may have bruises on their *rostral / caudal* regions.

    12. _caudal_

13. The external ear is *lateral / medial* to the inner ear.

    13. _lateral_

14. The fingers are *distal / proximal* to the wrist which is *distal / proximal* to the elbow which is *distal / proximal* to the shoulder.

    14. _distal_
        distal
        distal

15. The tongue is *superior / inferior* to the palate.

    15. _superior_

16. The scalp is *superficial / deep* in relation to the skull.

    16. _superficial_

17. The outer opening of the ear canal is *superior / inferior* to the ear lobe.

    17. _superior_

Note: If you can use and understand anatomical terms easily and automatically, you will be able to develop visual images of anatomical structures as you read the descriptions in the text. As you study anatomical drawings remember that the anatomical right is on the left side of the page.

Label the arrows in each drawing.

1. Superior—Inferior

2. Anterior—Posterior

3. Dorsal—Ventral

4. Proximal—Distal

5. Medial—Lateral

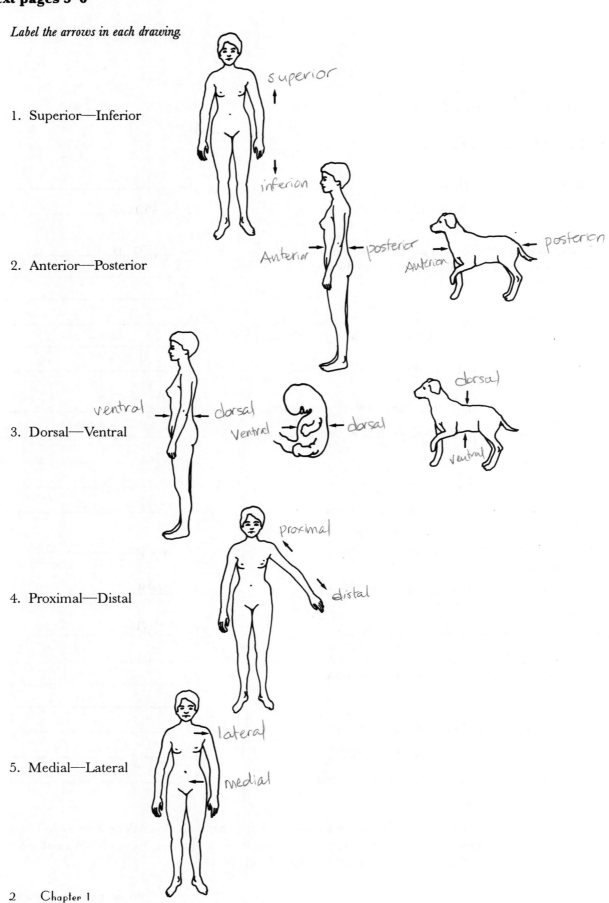

No. 1-2 cont'd

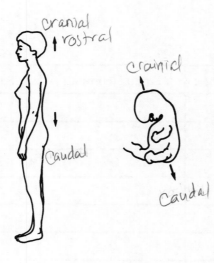

6. Cranial—Caudal
        or
    Rostral

7. External—Internal

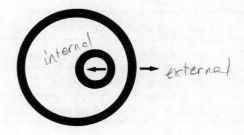

8. Superficial—Deep

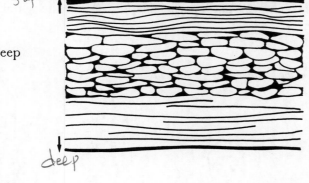

9. Peripheral—Central

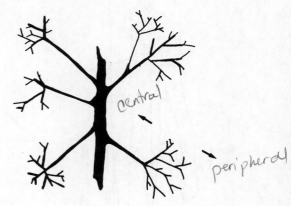

Word Association:

       The speaker's <u>rostrum</u> is at the <u>head</u> table.
       When you en<u>dorse</u> a check you sign it on the <u>back</u>.
       When you en<u>dorse</u> a political candidate you <u>back</u> a candidate.
       <u>Ventriloquism</u> (L. venter, belly + loqui, speak) <u>belly</u>-speech.
       <u>Proximity</u>; <u>near</u>ness
       Ap<u>proxim</u>ate; come <u>near</u> to
       <u>Distant</u>; <u>far</u>

Description Planes:	Frontal (coronal)	Sagittal	Transverse
cross-section			✔
longitudinal	✔	✔	
vertical	✔	✔	
horizontal			✔
divides the body into right and left halves		✔	
divides the body into anterior and posterior parts	✔		
divides the body into upper and lower parts			✔
derived from a word meaning crown	✔		
derived from a word meaning arrow		✔	

Question: Imagine two transverse sections that divide the torso into three equal parts. Now imagine a transverse plane in the middle section. What superficial landmark does a sagittal plane on the anterior or ventral wall bisect?

Label the planes on the illustrations below.

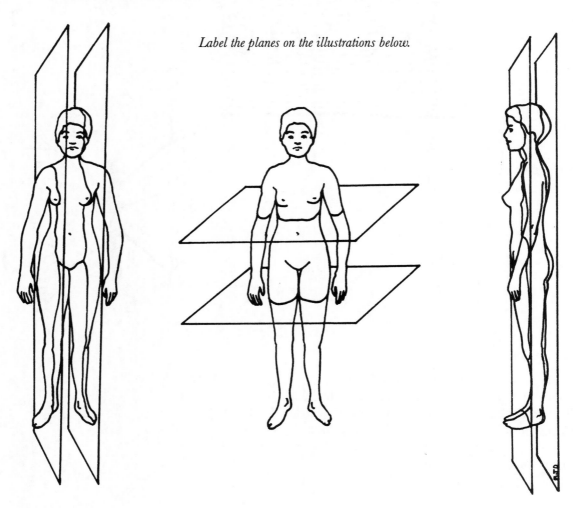

FIGURE 1.1 PICTORIAL SUMMARY OF PLANES AND GENERAL TERMS.

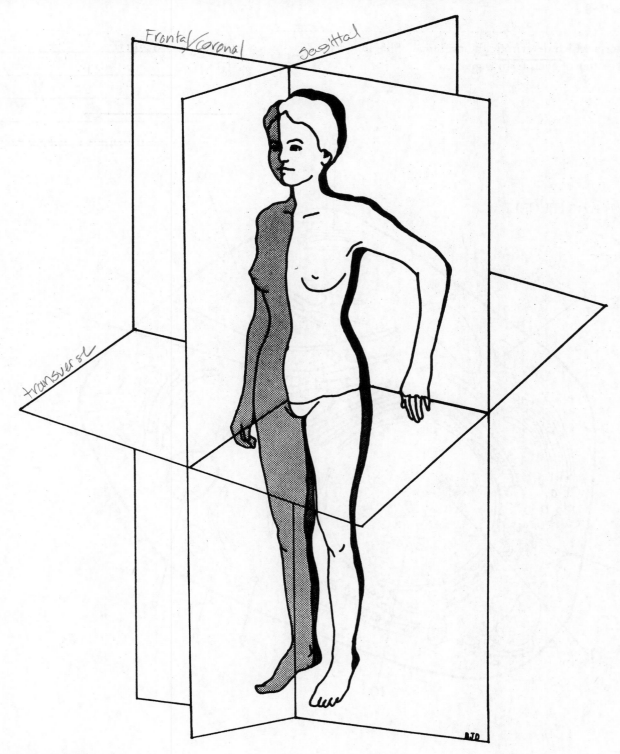

Locate and label:

1. frontal plane
2. sagittal plane
3. transverse plane
4. anterior
5. posterior

6. medial
7. lateral
8. ventral
9. dorsal
10. right

11. left
12. caudal
13. cranial
14. rostral

15. proximal
16. distal
17. dextrad
18. sinistrad

List criteria which determine the presence of life in a cell.
The first letter of each term is given.

g _growth_
r _reproduction_
i _irritable_
m _metabolism_
s _spontaneous movement_

FIGURE 1.2 THE CELL.

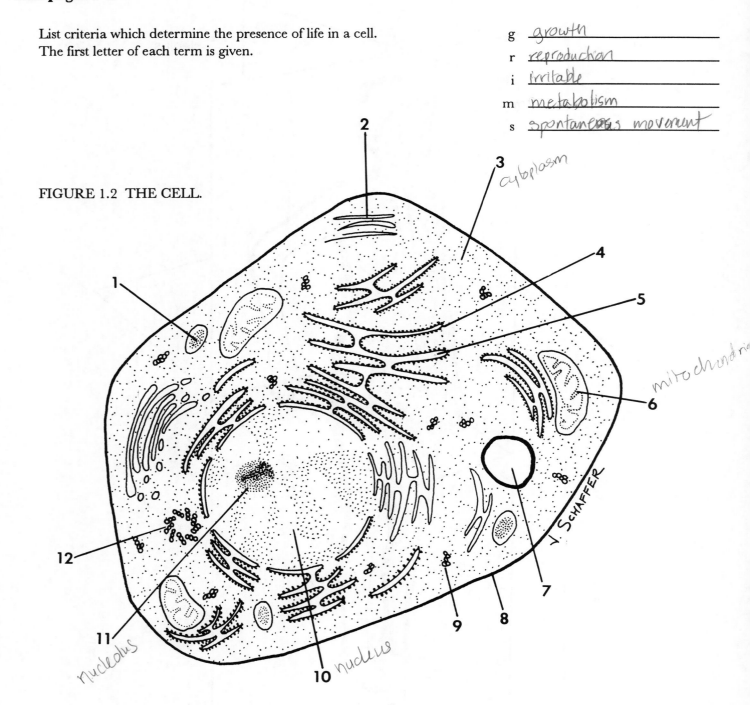

Identify:

1. lysosome
2. Golgi apparatus
3. cytoplasm
4. ribosome
5. endoplasmic reticulum
6. mitochondrion

7. vacuole
8. cell membrane
9. inclusion body
10. nucleus
11. nucleolus
12. centrosome

No. 1-5 CELLS (VOCABULARY)
Text pages 7-8

cell	DNA	nucleolus
cell membrane	endoplasmic reticulum	nucleus
centrosome	Golgi apparatus	protoplasm
cytology	lysosome	tissue
cytoplasm	mitochondria	vacuole

1. digestive organs of the cell

2. threadlike granules which provide energy in the form of ATP

3. body's fundamental unit of structure and function

4. located within the nucleus

5. contains ribosomes which are essential for protein synthesis

6. the study of cells

7. the mass of protoplasm around the nucleus

8. appears to temporarily store secretory substances

9. a colony of cells similar in structure and function

10. a chromatin deposit responsible for transmission of genetic traits

11. derived from words meaning central + body

12. a small cavity in the protoplasm of the cell

13. basic material of cell composition

14. forms an intracellular transport network

15. contributes to the formation of chromosomes during cell composition

16. controls the exchange of molecules and ions between the cell and its environment

17. contains DNA in its ground substance

1. _____

2. _____

3. _____

4. _____

5. _____

6. _____

7. _____

8. _____

9. _____

10. _____

11. _____

12. _____

13. _____

14. _____

15. _____

16. _____

17. _____

No. 1-6 ELEMENTARY TISSUES: OUTLINE
Text pages 8–29

Embryonic derivation of tissues is shown on text page 512, the cover page of Chapter 7.

TISSUE (Examples)	EMBRYONIC DERIVATION	DESCRIPTION	FUNCTION
Epithelial glands skin membranous linings	endoderm mesoderm ectoderm	• very little intercellular substance • free surface • rests upon a layer of connective tissue • single or multiple layers	protection secretion absorption glandular sensory
Connective tendons ligaments fascial membranes cartilage bone	mesoderm	• relatively few cells • large amount of intercellular substance (semifluid, fibrous or rigid)	support protection connection insulation food storage
Muscular	ectoderm	• elongated cells specialized for contraction	movement maintain posture
Nervous	mesoderm	• elongated cells specialized for irritability	reception conduction communication coordination integration
Vascular* blood lymph		• fluid intercellular substance • free cells (corpuscles)	transport and deliver oxygen, nutrients, and hormones to cells remove waste products and toxins from the body control temperature

*Vascular tissue is sometimes categorized as connective tissue.

Note: The substance of an organ is sometimes referred to as tissue, e.g., lung tissue, kidney tissue, but these are not single elementary tissues. Organs are usually composed of several types of tissue.

No. 1-7 EPITHELIAL TISSUES (CLASSIFICATION)
Text pages 9–10

Description Epithelial Tissues:	Epithelial Proper	Mesothelial	Endothelial
inner lining of blood and lymph vessels			✔
outer layer of skin and internal membranes which are continuous with the skin	✔		
lining of primary body cavities		✔	
some cells specialized to serve as sensory cells in exteroceptors, e.g. eye, ear	✔		
serous membrane		✔	
flat, cuboidal, or columnar cells in single or stratified layers	✔		
flat simple squamous cells in a single layer		✔	✔
peritoneal, pleural, and pericardial membranes		✔	

Define:

1. goblet cell _____

2. squamous _____

No. 1-8 EPITHELIAL TISSUE: SEROUS MEMBRANES LINING BODY CAVITIES
Text page 10

Description Serous Membranes:	Pleural	Pericardial	Peritoneal
derived from words meaning around + heart		✔	
derived from words meaning around + stretched			✔
mesothelial tissue	✔	✔	✔
lines the abdominal cavity			✔
lines the two lung cavities	✔		
lines the cavity containing the heart		✔	
a flat, single layer of cells resting on a sheet of loose connective tissue	✔	✔	✔

Note: *Pleurisy* or *pleuritis* is an inflammation of the pleural membranes. The word-ending *-itis* indicates inflammation. *Pericarditis* and *peritonitis* are then self-explanatory.

No. 1-9 CONNECTIVE TISSUE (CLASSIFICATION)
Text pages 10–16

Description Connective Tissue:	Loose	Dense	Special — Cartilage	Special — Bone
tightly packed fibers		✔	✔	✔
scattered fibers	✔			
semifluid intercellular substance	✔	✔		
intercellular substance firm but flexible			✔	
intercellular deposits of inorganic salts				✔
cells are called osteocytes and osteoblasts				✔
cells are called chondroblasts			✔	

No. 1-10 LOOSE CONNECTIVE TISSUE (CLASSIFICATION)
Text pages 10–11

Description Loose:	Areolar	Adipose
derived from a word meaning fatty		✔
derived from a word meaning space	✔	
found in subcutaneous fascia		✔
meshlike	✔	
has large, spherical cells		✔
forms "bed" for epithelial tissue	✔	

No. 1-11 DENSE CONNECTIVE TISSUE (CLASSIFICATION)
Text pages 11–12

Description Dense:	Tendons	Ligaments	Fascia	Reticular
attach bone to bone		✔		
attach muscle to bone	✔			
attach muscles to other muscles	✔			
separate and cover muscle fibers to form functional units			✔	
attach cartilage to cartilage		✔		
attach muscle to cartilage	✔			
attach bone to cartilage		✔		
feltlike				✔
support essential elements of organs				✔

No. 1-12 SPECIAL CONNECTIVE TISSUE: CARTILAGE
Text pages 12–13

Description Cartilage:	Hyaline	Elastic	Fibrous
probably does not calcify		✔	
may calcify or ossify with age	✔		
dense network of collagenous fiber and cartilage cells			✔
like milk glass, but yellows with age	✔		
covers articular surfaces of joints	✔		
forms intervertebral discs			✔
found in epiglottis and ear canals		✔	
forms framework of lower respiratory tract	✔		
found in small cartilages of larynx		✔	

1. Cartilage is more predominant in *youngsters / adults*.   1. _____
   Why? _____

2. The combining form meaning cartilage is *chondro / osteo*.   2. _____

No. 1-13 SPECIAL CONNECTIVE TISSUE: BONE
Text pages 13–15

Description Bone:	Compact (Dense)	Spongy (Cancellous)
interior of bone		✔
outer shell of bone	✔	
contain osteoblasts and osteocytes	✔	✔
consist of intersecting trabeculae		✔
pierced by Haversian canals	✔	
large quantity of inorganic salts deposited in matrix	✔	✔
appear solid to unaided eye	✔	
appear porous to unaided eye		✔
contains marrow (soft tissue of bone)		✔

No. 1-14 BONES (CLASSIFICATION)
Text page 14

accessory	long
air-containing	sesamoid
flat	short
irregular	

1. hip bones

2. length greater than width

3. vertebrae

4. bony structures within tendons

5. site of sinuses

6. leg bone

7. usually cuboidal, having several articular surfaces

8. having a supposed resemblance to sesame seeds

9. inner and outer plate of compact bone separated by a thin marrow space, as in the skull

10. Wormian bones, found between suture line in skull

11. found in ankle and wrist

1. _____

2. _____

3. _____

4. _____

5. _____

6. _____

7. _____

8. _____

9. _____

10. _____

11. _____

No. 1-15 BONES: AXIAL AND APPENDICULAR SKELETON
Text pages 14–15

Description/Components Skeleton:	Axial	Appendicular
derived from a word meaning to hang upon		✔
derived from a word meaning central line	✔	
pelvic girdle		✔
rib cage	✔	
hyoid bone	✔	
leg		✔
shoulder girdle		✔
skull	✔	
mandible (lower jaw)	✔	
arm		✔
vertebral column	✔	
more important in speech and hearing science	✔	

No. 1-16 SURFACE STRUCTURE OF BONE: ELEVATIONS AND DEPRESSIONS
Text pages 15-16

Elevations or Prominences

condyle   spine
crest   trochanter
head   tubercle
process   tuberosity

1. a narrow ridge of bone

2. a general term for bony prominence

3. a rounded or knucklelike eminence which articulates with another bone

4. a small rounded projection

5. an enlargement at one end of a bone beyond the constricted portion

6. a large round projection

7. a sharp projection

8. a very large bony projection

1. _____

2. _____

3. _____

4. _____

5. _____

6. _____

7. _____

8. _____

Depressions or Cavities

fissure   meatus
foramen   neck
fossa   sinus
fovea   sulcus

1. a cleft or deep groove

2. a cavity within a bone

3. an opening or perforation in a bone

4. a constriction near one end of a bone

5. a hollow or shallow depression

6. a groove or furrow

7. a tubelike passageway

8. a small pitlike depression

1. _____

2. _____

3. _____

4. _____

5. _____

6. _____

7. _____

8. _____

anastomose	diaphysis	osteocytes
aponeuroses	diploe	parenchyma
appositional growth	epiphysis	perichondrium
bone marrow	interstitial growth	periosteum
canaliculi	lacunae	red marrow
chondroblasts	lamellae	subcutaneous fascia
collagenous	matrix	trabeculae
cortex	osteoblasts	yellow marrow

1. delicate spicules of bone which intersect each other to form a very complicated meshwork

2. to interconnect parts of a branching system

3. irregular spaces

4. smooth shaft of long bone

5. cells of cartilage

6. soft tissue that fills cavity of bone

7. tough, fibrous membrane investing cartilage

8. small canals

9. broad tendinous sheets

10. bone-forming cells

11. noncellular components of connective tissue

12. thin layers

13. yields gelatin when boiled in water

14. marrow space in skull

15. expansion due to cell multiplication

16. essential or functional elements of an organ

17. outer shell

18. pure adipose tissue

19. tough, fibrous membrane investing bone

20. a continuous sheet of dense connective tissue covering the entire body; located between skin and deeper structures

21. a bone cell embedded in a rigid matrix of bone

22. due to deposition or formation at periphery

23. manufactures red blood cells

24. head of the bone, an articular facet or condyle

1. _____

2. _____

3. _____

4. _____

5. _____

6. _____

7. _____

8. _____

9. _____

10. _____

11. _____

12. _____

13. _____

14. _____

15. _____

16. _____

17. _____

18. _____

19. _____

20. _____

21. _____

22. _____

23. _____

24. _____

No. 1-18 JOINTS (FUNCTIONAL CLASSIFICATION)
Text pages 16–18

Description Joints:	Synarthrodial	Amphiarthrodial	Diarthrodial
immovable	✔		
freely movable			✔
slightly movable or yielding		✔	
derived from words meaning double + joint			✔
derived from words meaning on both sides + joint		✔	
derived from words meaning together + joint	✔		
cartilaginous joint		✔	
synovial joint			✔
fibrous joint	✔		
bones are almost in direct contact	✔		
bones are joined by a band of fibrous tissue which forms the articular capsule			✔
bones are joined by a thin, intervening tissue	✔		
found in skull	✔		
contiguous bone edges are united by cartilage		✔	
lubricated by synovial fluid			✔

Note: Synovial is derived from words meaning together + egg. Synovial fluid has the consistency of raw egg white.

No. 1-19 SYNARTHRODIAL JOINTS (CLASSIFICATION)
Text pages 16–17

Description Synarthroidal:	Suture	Schindylesis	Gomphosis	Syndesmosis
a single plate of bone is inserted into a cleft which has been formed by the separation of two laminae in another bone		✔		
found in the middle ear				✔
found in the skull	✔	✔	✔	✔
derived from a word meaning seam	✔			
articulation of a conical process into a socket			✔	
bones which are almost touching are joined together by a thin, intervening tissue	✔	✔	✔	✔
two bones united by interosseous ligaments				✔
roots of teeth inserted in their sockets			✔	
restrict or prevent movement	✔	✔	✔	✔
fibrous joint	✔	✔	✔	✔

Describe:

1. serrated suture _____

2. dentate suture _____

3. sutura limbosa _____

Note:

syn-	signifies union or association
ost-	derives from a word meaning bone
dys-	signifies disordered, difficult, bad, painful
-osis	denotes a process, often a disease or an abnormal process
synostosis	a union between adjacent bones or parts of a single bone formed by osseous material
dysostosis	defective ossification; defect in normal ossification of fetal cartilages
syndrome	a set of symptoms which occur together

The names of the following syndromes should thus become more meaningful:

> orodigitofacial dysostosis
> dysostosis multiplex (Hurler's syndrome)
> craniofacial dysostosis (Crouzon's syndrome)
> mandibulofacial dysostosis (complete: Franceschetti's syndrome)
> mandibulofacial dysostosis (incomplete: Treacher Collins syndrome)

No. 1-20 AMPHIARTHRODIAL JOINTS (CLASSIFICATION)
Text page 17

Description Amphiarthroidal:	Synchondrosis	Symphysis
cartilaginous	✔	✔
yielding	✔	✔
allows for growth of the skull and long bones	✔	
surfaces of bone which are covered with hyaline cartilage are connected by discs of fibrocartilage		✔
because the joint eventually ossifies, it is only a temporary cartilaginous union	✔	
broadly represented in skeleton	✔	✔
premature ossification of the joint may be associated with craniofacial dysostosis and mental retardation	✔	

Note:

arthro- *	derived from a word meaning joint
chondro-	derived from a word meaning cartilage
-itis	denoting inflammation

    * to be differentiated from arthria, (Gk. arthoun, to utter distinctly) as in dysarthria.

    Analyze each word and draw a line to its correct definition.

arthritis	inflammation of cartilage
arthrochondritis	inflammation of the cartilage in a joint
chondritis	inflammation of bone and cartilage
osteochondritis	inflammation of a joint

ball and socket
condyloid
gliding

hinge
pivot
saddle

1.

2.

1. _____

2. _____

3.

4.

3. _____

4. _____

5.

6.

5. _____

6. _____

7. permit rotation (2)

8. permit all types of motion except rotation (2)

7. _____

_____

8. _____

_____

Description/Illustration	Striated	Smooth
	✔	
		✔
innervated by automatic nervous system		✔
long fibers	✔	
spindle-shaped cells		✔
involuntary muscle		✔
voluntary muscle	✔	
visceral muscle		✔
skeletal muscle	✔	
more primitive		✔

Cardiac Muscle

1. has properties of *smooth muscle / striated muscle / both.*   1. _____

2. is *voluntary / involuntary / both.*   2. _____

3. is *striated / unstriated / both.*   3. _____

No. 1-23 MUSCLE TISSUE (VOCABULARY)
Text pages 18–20

endomysium   fusiform   myoglobin
ephaptic conduction   kinesiology   perimysium
epimysium   myocardium   sarcolemma
fascia   myofibril   sarcoplasm
fasciculi

1. a muscle filament contained in a muscle fiber

2. a fibrous intermuscular septum separating muscle groups

3. cardiac muscle

4. specialized protoplasm in which myofibrils are embedded

5. spindle-shaped cells contained in smooth muscle

6. a moderately coarse fibrous tissue ensheathing fasciculi

7. the science of movement

8. fibrous tissue which binds muscle fibers and separates them from adjacent muscle fibers

9. delicate elastic, transparent and homogenous membrane which invests every striated muscle fiber

10. groups of muscle fibers

11. non-neural transmission of contractile impulse to adjacent fibers resulting in wavelike contraction over entire muscular organ

12. a coarse fibrous tissue encasing an entire muscle

13. a protein which increases diffusion of oxygen into muscle fibers and contributes to their color

1. _____

2. _____

3. _____

4. _____

5. _____

6. _____

7. _____

8. _____

9. _____

10. _____

11. _____

12. _____

13. _____

No. 1-24 MUSCLE CONTRACTION (CHARACTERISTICS)
Text pages 20–23 *(does not cover supplemental notes*)*

1. The maximum force of contraction is generated when a muscle is:
   a. at normal resting length
   b. stretched beyond resting length
   c. shorter than resting length

1. _____

2. Under ideal conditions a muscle may shorten as much as:
   a. 10-20 percent
   b. 30-40 percent
   c. 50-60 percent

2. _____

3. The maximum strength of muscle contraction per square centimeter is:
   a. 3 kilograms
   b. 5 kilograms
   c. 7 kilograms

3. _____

* small print

4. When a muscle has become moderately fatigued: 4. _____

    a. nerve impulses subside   _____

    b. contractions weaken

    c. metabolic processes cannot provide enough energy for
        muscle cells

5. When a muscle becomes excessively fatigued: 5. _____

    a. it will not contract   _____

    b. it may stay contracted and rigid for several minutes

    c. ATP is depleted

    d. actin and myosin filaments separate

Note: Physicians may use electromyography (EMG) to distinguish between muscle and nerve pathology, and to
map the site of the defect.

No. 1-25 MUSCLE CONTRACTION (VOCABULARY)
Text pages 20–24

actin filament   isotonic contraction
ATP (adenosine triphosphate)   muscle tone
electrocardiogram   myosin filament
electromyography   rigor mortis
endoplasmic reticulum   sarcomere
isometric contraction   single muscle twitch

1. EMG recording of bioelectric activity due to contraction   1. _____
    of the heart muscle

2. muscle shortens but tension remains constant   2. _____

3. graphically recording bioelectric activity due to   3. _____
    muscle contraction

4. muscle contraction which takes place several hours   4. _____
    after death

5. the segmented tubular sleeve surrounding each myofibril   5. _____

6. muscle tenses but does not shorten   6. _____

7. stimulated by very short-duration excitation to the nerve   7. _____
    or muscle; means of studying muscle contraction

8. individual contractile unit composed of actin and myosin   8. _____
    filaments

9. a protein molecule which is located within a myofibril and   9. _____
    which appears light in color when under polarized light

10. a long protein molecule which is located within a myofibril   10. _____
    and which appears dark in color when under polarized light

11. slight contractile tension remaining when a muscle is   11. _____
    "at rest"

12. nourishes the actin and myosin filaments   12. _____

No. 1-26 MUSCLE ARCHITECTURE (CLASSIFICATION)
Text pages 24–25

Description/Illustration	Parallel	Pennate	Radiating
derived from a word meaning feather		✔	
has great range of motion but limited power	✔		
fasciculi converge from a broad surface to a narrow point			✔
fasciculi are parallel to the long axis of the muscle	✔		
fasciculi converge obliquely along the length of the tendon		✔	
two types of muscles having a lesser range of motion, but greater power		✔	✔
			✔
	✔		
		✔	

No. 1-27 MUSCLE ATTACHMENT (TYPES)
Text page 25

Description	Origin	Insertion
attachment which is usually fixed or engages in the lesser movement	✔	
structure acted upon		✔
in the extremities: most proximal	✔	
in the extremities: most distal		✔

antagonist   mechanical advantage
applied force   mechanical disadvantage
fixation muscle   prime mover
fulcrum   resistance force
gravity   synergist
lever arm

1. the force against which postural muscles usually work   1. _____

2. a muscle which protects a joint during extreme motion and which may function as a synergist   2. _____

3. a muscle directly responsible for producing desired movement   3. _____

4. a rigid bar or plate acted upon at different points by two forces which tend to rotate it about a fixed axis or fulcrum   4. _____

5. the power employed in a lever system   5. _____

6. a muscle which suppresses undesired action; derived from words meaning together + work   6. _____

7. a large applied force is required to move a small resistance force   7. _____

8. functions to maintain appropriate posture   8. _____

9. a small applied force is required to move a large resistance force   9. _____

10. the support or point on which a lever rotates   10. _____

11. the weight to be moved by leverage   11. _____

The Biological Lever System

applied force
fulcrum
lever arm

(1) contracting skeletal muscles   (1) _____

(2) bone   (2) _____

(3) joint   (3) _____

Description	Mech. Advantage or Disadvantage		Class		
	Adv. (Disadv.)		(1)	2	3
	(Adv.) Disadv.		1	(2)	3
	(Adv.) Disadv.		(1)	2	3
	Adv. (Disadv.)		1	2	(3)
exemplified by wheelbarrow			1	(2)	3
exemplified by teeter-totter			(1)	2	3
exemplified by drawbridge			1	2	(3)
sacrifices power for speed			1	2	(3)
most common in the body			1	2	(3)

Label each drawing.

1. Flexion—Extension

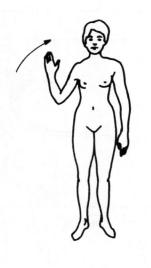

2. Abduction—Adduction

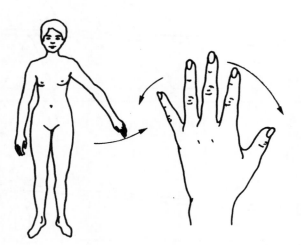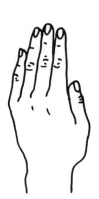

3. Supination—Pronation

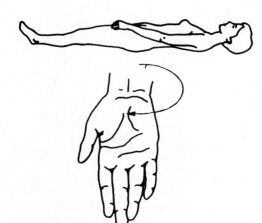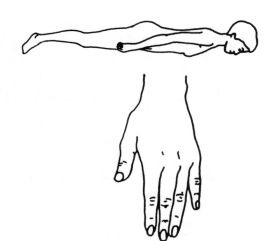

Attempt to demonstrate the following actions and positions. You may refer to the definitions and illustrations in the text and study guide to determine how well you have followed the instructions.

1. Assume the anatomical position.
2. Flex your hand.
3. Extend your hand.
4. Oppose the thumb and little finger of your right hand.
5. Extend your arm.
6. Evert your left foot.
7. Pronate your left hand.
8. Abduct your fingers.
9. Circumduct your head.
10. Rest supine on the floor.
11. Demonstrate dorsiflexion (backward flexion or bending) of the foot.
12. Demonstrate plantar flexion of the foot.
13. Pronate your right hand.
14. Adduct your fingers.
15. Extend your neck.
16. Evert your right foot.
17. Invert your left foot.
18. Abduct your vocal folds (by taking a deep breath).
19. Oppose the thumb and index finger of your left hand.
20. Flex your arm.
21. Medially rotate your right leg.
22. Supinate your left hand.
23. Flex your neck.
24. Rest prone on the floor.
25. Adduct your vocal folds (by pretending you are going to cough).
26. Extend your hand.
27. Laterally rotate your left leg.
28. Abduct your arms.
29. Adduct your arms.

Note: A *plantar reflex* (plantar flexion of the foot) is a normal response to stimulation of the sole of the foot.

Babinski's reflex (dorsiflexion of the big toe) is an abnormal response to stimulation of the sole of the foot. It may be indicative of lesions in the pyramidal tract of the nervous system and is often associated with spasticity in cerebral palsy.

Suggestion: By increasing your awareness of motor problems and by testing your own ability to identify and describe them, you will be developing an important professional skill. Something as seemingly simple as the client's *right* and *left* may be a bit confusing at first.

attachments (origin-insertion)   geometric shape
function   location in body
general form   number of heads at origin

Name of Muscle | Information Contained in Name (one answer)

1. subclavius   1. _____
2. intercostal   2. _____
3. levator costalis   3. _____
4. triangularis   4. _____
5. longitudinal   5. _____
6. constrictor   6. _____
7. tensor palatini   7. _____
8. palatopharyngeal   8. _____
9. biceps   9. _____
10. thyrohyoid   10. _____
11. nasalis   11. _____
12. quadriceps   12. _____
13. depressor septi   13. _____
14. serratus   14. _____

Other Descriptive Terms (most are self-explanatory)   Prefixes

superior	— inferior	rectus (straight)	infra-
anterior	— posterior	transverse	supra-
lateral	— medial	oblique	sub-
major	— minor		inter-
brevis	— longus		
maximus	— minimus		
dorsalis	— ventralis		

Note: In nautical terminology *stern* means back, but in anatomical terminology *sterno-* refers to the sternum (breastbone).

Suggestion: When you first encounter the name of a muscle, see how much information you can derive from its name.

1. Nerve cells (neurons) are *short / long*.

2. Nerve cells are specialized for *irritability / contraction*.

3. If you should prick your finger, the nerve cells in that area will modify their *electrochemical composition / physical size / both*.

1. _____

2. _____

3. _____

FIGURE 1.3 SCHEMATIC OF A MOTOR UNIT.

Identify:

1. dendrite
2. cell body
3. cell nucleus
4. axon
5. axon collateral
6. motor end plates
7. muscle fibers

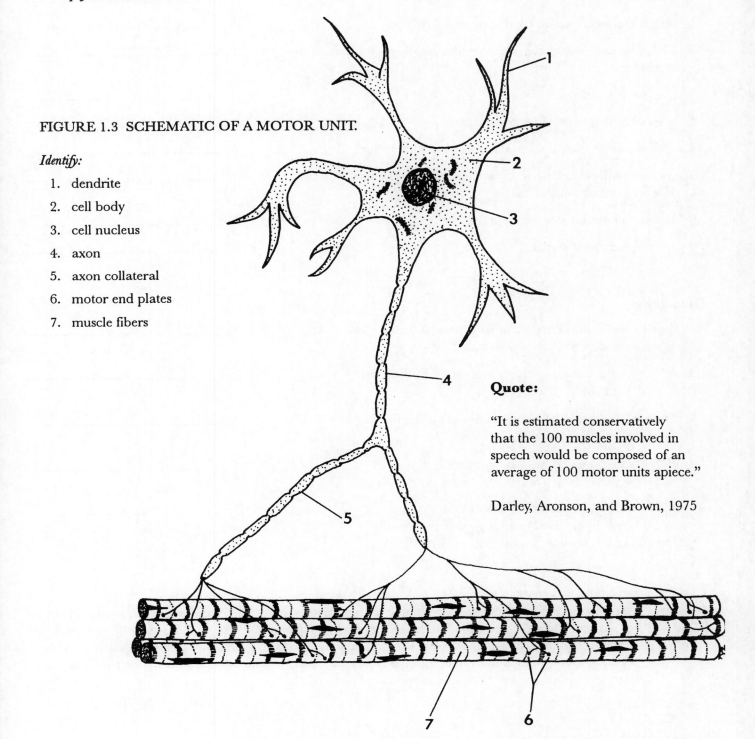

Quote:

"It is estimated conservatively that the 100 muscles involved in speech would be composed of an average of 100 motor units apiece."

Darley, Aronson, and Brown, 1975

axon
contraction period
latent period
motor unit

muscle end plate
relaxation period
refractory period
sarcoplasm

1. interval between onset of stimulus and onset of contraction

1. _____

2. functional structure for producing muscle action

2. _____

3. chemical processes occur which restore muscle to normal resting state

3. _____

4. a nerve cell process

4. _____

5. passive return to previous relaxed state

5. _____

6. transmits nerve impulses to sarcoplasm of muscle fiber

6. _____

7. includes a nerve cell body, its processes, and all muscle fibers served by a nerve cell

7. _____

8. longitudinal substance between muscle fibrils

8. _____

9. muscle is actively working

9. _____

Questions:

(1) In a single motor unit will a strong stimulus activate more muscle fibers than a weak stimulus?

(1) _____

(2) Will a strong stimulus activate more motor units than a weak stimulus?

(2) _____

(3) Does the frequency of stimulus impulses affect the forces exerted by muscle fibers?

(3) _____

(4) Will greater force be exerted by a muscle if more of its motor units are activated?

(4) _____

(5) In general, would the motor units of the speech musculature be smaller than the motor units for the muscles of locomotion or posture?

(5) _____

Quote: "For the most part animals receive stimuli from external sources and respond to them principally by movement. Between perception and performance lies the intricate system of nerves and nerve centers which coordinate the whole and make orientation and directive action possible. This reaction system, operating in conjunction with an internal biological clock, enables animals to be in the right place at the right time—the key to success in most vital matters."

N.J. Berrill, 1966

No. 1-35 VASCULAR TISSUE (VOCABULARY)
Text page 29

blood	blood plasma	blood platelets	erythrocytes
leukocytes	lymph	lymphocytes	thromboplastin

1. an enzyme which contributes to the clotting of the blood   1. _____
2. red blood cells   2. _____
3. lymph cells   3. _____
4. corpuscles and platelets separated by plasma   4. _____
5. protoplasmic dishes   5. _____
6. fluid intercellular portion of blood   6. _____
7. white blood cells   7. _____
8. immediate nutrient plasma of tissues   8. _____

Review the functions of vascular tissue.

No. 1-36 ORGANS
Text page 29

1. A somewhat independent part of the body that performs a special function is called a/an _____.   1. _____
2. The cells that compose the functional elements of an organ are called _____.   2. _____
3. Does an organ generally contain several types of tissue?   3. _____
4. Is one kind of tissue usually dominant in an organ?   4. _____
5. What is the primary respiratory organ?   5. _____
6. What is the organ of hearing?   6. _____
7. What is the organ of the voice?   7. _____

Note: Smaller organs may be a rather independent part of larger organs. For example, the tastebuds (sensory end organs) are on the tongue which is an important organ of mastication and speech.

No. 1-37 SYSTEMS
Text page 30

A *system* is an arrangement of related organs exhibiting functional unity.

In the following exercise fill in the name of the system and the field of study associated with that system. The first letter of each answer is given.

Components of the System	Name of the System	Field of Study
1. bone, cartilage	S _____	O _____
2. joints, ligaments	A _____	A _____
3. muscles	M _____	M _____
4. digestive tract, glands	D _____	S _____
5. heart, blood, lymph	V _____	A _____
6. brain, spinal cord, nerves, sense organs	N _____	N _____
7. lungs, air passageways	R _____	P _____
8. kidneys, urinary passageways	U _____	U _____
9. genital tract	G _____	G _____
	or	(of female)
	R _____	
10. ductless glands	E _____	E _____
11. skin, nails, hair	I _____	D _____

No. 1-38 BASIC STRUCTURE (REVIEW)
Text pages 7-30

cell	gland	organ	tissue	system

1. a highly organized mass of protoplasm which possesses life
2. two or more organs combined to exhibit functional unity
3. a cell, tissue, or organ which produces and discharges a substance used elsewhere in the body
4. two or more tissues combined to exhibit functional unity
5. a colony of cells similar in structure and function
6. the brain
7. an osteoblast
8. serous membrane
9. the heart
10. epidermis
11. a muscle fiber
12. the eye
13. the thyroid gland (2)

14. skin
15. blood
16. a leukocyte
17. epithelium
18. an aponeurosis
19. a neuron
20. smooth muscle

1. _____
2. _____
3. _____
4. _____
5. _____
6. _____
7. _____
8. _____
9. _____
10. _____
11. _____
12. _____
13. _____
_____
14. _____
15. _____
16. _____
17. _____
18. _____
19. _____
20. _____

No. 1-39 SPEECH PRODUCTION
Text pages 30-32

Physical Analog of Speech Mechanism	Phase of Speech Production	Structures Involved
1. power supply	_____	_____
2. vibrating elements	_____	_____
3. system of valves	_____	_____
4. system of filters	_____	_____

What are the shortcomings of describing speech production in terms of "phases"?

_____

_____

_____

Questions: Why are the concepts of temporal overlap and mutual influence particularly important to your understanding of speech production?

How would you explain speech production to a first-grade class?

Chapter 2
Breathing

No. 2-1 BREATHING: DEFINITION
Text page 34

chemical
mechanical
physical

1. The gas exchange between an organism and its environment   1. _____
   is a _____ process.

2. The inhalation and exhalation of air, without concern for   2. _____
   respiratory function, is a _____ process.

3. The oxidation of food to produce water, carbon dioxide,   3. _____
   and heat is a _____ process.

No. 2-2 THE PHYSICS OF BREATHING: THE KINETIC THEORY OF GASES
Text pages 34–35

1. Kinetic is derived from a word meaning _____.   1. _____

2. Unceasing free movement of molecules is most prominent   2. _____
   in *solids / liquids / gases*.

3. The force exerted on the walls of a gas-filled container   3. _____
   is a function of the number of gas molecules within
   the vessel, provided volume and _____ are held
   constant.

4. Molecules exert a force only when they collide with   4. T   F
   something.

5. When a balloon is punctured only the molecules   5. T   F
   which randomly encounter the pinhole will flow to
   the outside.

Boyle's Law: If a gas is kept at constant temperature, its pressure and volume are inversely proportional. (If one goes up the other goes down by an equal proportion.)

1. pressure × volume = a constant   1. T F

2. $P \times V = K$ 2. T F

3. $\dfrac{P}{V} = K$ 3. T F

4. $P_1 V_1 = P_2 V_2$*   4. T F

If two factors are inversely proportional they have a constant product.

5. If $P \times V = 12$, then: 5.

    a. if $P = 3$, $V = $ _____   a. $V = $ _____

    b. if $V = 6$, $P = $ _____   b. $P = $ _____

    c. if $P = 12$, $V = $ _____   c. $V = $ _____

If two factors are inversely proportional the division of one factor by X will result in the multiplication of the other factor by X.

6. According to Boyle's Law: 6.

    a. $2P \times 2V = K$ a. T F

    b. $\dfrac{2P}{2V} = K$ b. T F

    c. $2P \times \dfrac{V}{2} = K$ c. T F

    d. $\dfrac{P}{3} \times 3V = K$ d. T F

    e. $2P \times \dfrac{V}{2} = 4P \times \dfrac{V}{4}$ e. T F

7. When the pressure compressing a gas is doubled, the   7. T F
    density of the gas is doubled.

---

     *$_1$ = initial state; $_2$ = final state

No. 2-4 THE PHYSICS OF BREATHING: USING BOYLE'S LAW
Text pages 34–35

Note: See text page 84 for an explanation of measurement of pressure using columns of mercury. Hg is the symbol for mercury.

A gram-molecule is the amount of substance with a mass of the number of grams of its molecular weight.

1. Air at atmospheric pressure is confined in an airtight container with a movable piston at the top. $P \times V = K$

   Pressure = 1 atmosphere (760 mm Hg)

   Volume = 1 gram-molecule of gas at 0° C.
   (22.4 liters)

   Differential pressure is *zero / positive / negative*.

   1. _____

2. The piston is pushed in until the volume is halved.

   $$2 \ (760 \text{ mm Hg}) \times \frac{22.4 \text{ liters}}{2} = K$$

   Pressure = _____

   Volume = _____

   Differential pressure is *zero / positive / negative*.

   2. P = _____

   V = _____

   _____

3. The piston is pushed in until the volume is again halved.

   $$X \ (760 \text{ mm Hg}) \times \frac{22.4 \text{ liters}}{X} = K$$

   Pressure = _____

   Volume = _____

   Differential pressure is *zero / positive / negative*.

   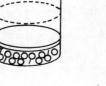

   3.

   X = _____

   P = _____

   V = _____

   _____

4. Usually an increase in the volume of the chest cavity results in a *negative / positive* pressure within the lungs with respect to atmospheric pressure.

   Air will rush *into / out of* the lungs.

   This is called *inhalation / exhalation*.

   4. _____

   _____

   _____

5. Usually a decrease in the volume of the chest cavity results in a _____ pressure within the lungs with respect to atmospheric pressure.

   Air will rush _____ the lungs.

   This is called _____.

   5. _____

   _____

   _____

6. The greater the difference between atmospheric pressure and the pressure within the lungs, the greater the volume of _____.

   6. _____

1. Label the schematic of the respiratory passage. Bracket the upper and lower respiratory tracts.

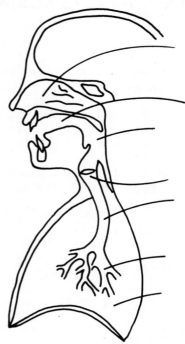

2. What are the anatomical terms for the following colloquial terms?

   a. mouth

   b. windpipe

   c. throat

   d. vocal cords

   e. voicebox

   f. roof of the mouth

3. What separates the upper and lower respiratory tracts?

4. How does the upper respiratory tract modify air which has been inhaled through the nose?

5. The laryngeal framework is a modification of the uppermost _____ cartilages.

6. A sudden release of compressed air by the larynx resulting in an explosive exhalation that will clear the passageway of accumulated mucus and foreign matter is called a/an _____.

7. The closure of the valvular laryngeal mechanism permits _____ fixation.

2.

   a. _____

   b. _____

   c. _____

   d. _____

   e. _____

   f. _____

3. _____

4. _____

    _____

    _____

5. _____

6. _____

7. _____

Note: The Heimlich maneuver, a sudden upward thrust on the upper abdomen, is a first-aid procedure used to dislodge the blockage when a person is choking. It is used only when someone is unable to cough, speak, or breathe.

Question: In terms of mechanics, why is the Heimlich maneuver effective?

1. The length of the trachea is approximately 11–12 centimeters or *4½ / 6½ / 9* inches.

2. The diameter of the trachea is approximately 2–2.5 centimeters or *1 / 2 / 3* inch(es).

3. The trachea is *anterior / posterior* to the esophagus.

4. The incomplete portions of the horseshoe-shaped tracheal rings are located *anteriorly / posteriorly*.

5. The tracheal rings are composed of _____ cartilage.

6. The first tracheal cartilage is connected to the cricoid cartilage of the larynx by means of the _____ ligament.

7. The last cartilage of the trachea divides or bifurcates to form the _____.

8. In the spaces between the tracheal rings the outer and inner tracheal membranes blend to form the _____ membrane.

9. The muscles in the spaces between the ends of the tracheal rings are *longitudinal / transverse / both*. These muscles are normally in a state of *contraction / relaxation*.

10. The trachea is lined with _____ membrane which is continuous with that of the larynx and bronchi.

11. The mucus is produced by mucous glands and _____ cells.

12. Threadlike cytoplasmic processes which beat rhythmically are called _____. They help clear the respiratory tract of contaminants and _____.

13. The surgical procedure which provides a temporary alternative airway is called a *tracheotomy / tracheostomy*.

14. The surgical procedure in which the superior border of the trachea is brought forward and sutured to the skin of the neck, thus forming a permanent airway is called a *tracheotomy / tracheostomy*.

1. _____

2. _____

3. _____

4. _____

5. _____

6. _____

7. _____

8. _____

9. _____

   _____

10. _____

11. _____

12. _____

    _____

13. _____

14. _____

Mucus is a noun, mucous an adjective.

Question: How does smoking affect ciliary action?

Note: Indications for Tracheotomy

Obstruction at or above the larynx which severely hinders breathing	Inability to cough-up secretions from the lower respiratory tract (in effect, drowning in mucus)
Examples: 1. cancer of the larynx 2. choking on food which cannot be dislodged 3. severe infections of the larynx or upper respiratory tract 4. injuries of the tongue, mandible, or larynx	Examples: 1. paralysis of muscles of respiration 2. pneumonia 3. unconsciousness 4. drug overdose 5. pain resulting from trauma (e.g. fractured ribs)
Symptoms of obstruction: 1. pallor 2. dyspnea (difficult, labored breathing) 3. inspiratory stridor (a high-pitched, harsh sound accompanying inhalation) 4. a sucking-in of muscles in the suprasternal notch at the base of the neck	

No. 2-7 THE RESPIRATORY PASSAGE: THE BRONCHI
Text pages 37–38

1. The bronchi are tubes which extend from the _____ to the _____.

     1. _____ _____

2. The hilum of the lung is at the entrance of the _____ into the lung.

     2. _____

3. The divisions of the bronchial tree:

     3.

      a. main or _____

       a. _____

      b. lobar or _____

       b. _____

      c. segmental or _____

       c. _____

4. The construction of the bronchi is much like that of the _____.

     4. _____

5. Inflammation of the bronchi is called _____.

     5. _____

6. Why are foreign bodies more likely to enter the right bronchus? _____

_____

No. 2-7 cont'd

Characteristics	Left Bronchus	Right Bronchus
supplies the larger lung		✔
is longer	✔	
is larger in diameter		✔
divides into two secondary bronchi, one for each lobe of the lung, and then into eight tertiary bronchi	✔	
divides into three secondary bronchi, one for each lobe of the lung, and then into ten tertiary bronchi		✔

Note: Symptoms of a foreign body in the bronchus:

1. difficult labored breathing, probably becoming progressively worse

2. wheezing during both inhalation and exhalation

3. recurring periods of violent coughing

A *bronchoscope* is an instrument for examining the interior of the bronchi and for removing foreign bodies from them.

No. 2-8 THE RESPIRATORY PASSAGE: THE BRONCHIOLES AND ALVEOLI
Text page 38

1. The tertiary bronchi repeatedly subdivide until they are almost microscopic, and finally give rise to the _____.

1. _____

2. In the bronchial tree the combined cross-sectional area of any given subdivision is *lesser / greater* than the cross-sectional area of the parent division.

2. _____

3. Repeated divisions of the bronchioles ultimately give rise to the _____ bronchioles which communicate directly with the _____ ducts that open into the _____ of the lung.

3. _____
_____
_____

4. As the bronchi and bronchioles divide, their structure becomes increasingly *muscular / cartilaginous.*

4. _____

5. Small depressions that pit the walls of terminal bronchioles and air sacs are called _____. Their walls are invested with a network of _____.

5. _____
_____

6. How is the rapid exchange of oxygen and carbon dioxide facilitated? _____

6. _____

1. The mediastinum contains blood and lymph vessels, nerves, the esophagus, and the _____.

   1. _____

2. Most of the elastic properties of the lung can be accounted for by *tissue elasticity / properties of the alveoli.*

   2. _____

3. The molecules of the moist pulmonary alveolar epithelium and of the air demonstrate a universal attraction at the air-liquid _____.

   3. _____

4. Because of surface tension, the alveoli tend to *expand / collapse.* The substance that reduces this tension is called _____.

   4. _____

   _____

5. A sewing needle will float on the surface of clean water due to a phenomenon called _____. If a few drops of detergent are added to the water the needle will sink. The detergent acts as a/an _____.

   5. _____

   _____

6. In proportion to the size of the thorax, an infant's lungs are *small / large.*

   6. _____

7. The bases of the lungs are separated from the abdominal viscera by the _____.

   7. _____

8. A young male adult may have a lung capacity in excess of 5,000 cubic centimeters or approximately _____ liters.

   8. _____

9. The parietal (costal) pleura lining the _____ cavity and the visceral pleura investing the _____ are anatomically contiguous membranes.

   9. _____

   _____

10. What is the intrapleural space? _____

11. What are the functions of the pleurae? _____

    _____

Description	Right Lung	Left Lung
smaller		✔
site of cardiac impression		✔
over the bulk of the liver	✔	
has two lobes		✔
has three lobes	✔	

Note: *Pneumothorax*, which is defined as an accumulation of gas or air within the pleural cavity, may occur

   (1) *spontaneously*, without known cause

   (2) *traumatically*, e.g., due to a perforating chest wound

   (3) *pathologically*, e.g., due to the rupture of a portion of a diseased lung

   (4) *artificially*, due to intentional introduction of air or gas into the pleural cavity, usually through a needle

No. 2-10 LUNGS (VOCABULARY)
Text pages 38–43

hilum
in situ
lung
mediastinum
pericardium
pleurae
pleural sinuses

pleurisy
pneumothorax
pulmonary ligament
pulmonary alveolar epithelium
surface tension
surfactant
thorax

1. a detergentlike substance which reduces surface tension.

1. _____

2. derived from a word meaning light in weight

2. _____

3. the presence of gas or air in the pleural cavity

3. _____

4. depression on the mediastinal surface of the lung where its root (bronchi, blood vessels, nerves) emerges

4. _____

5. the space in the central region of the thorax between the two pleura

5. _____

6. a property of liquid or solid matter which is the result of unbalanced molecular forces near the surface

6. _____

7. inflammation of the pleura

7. _____

8. serous membranes which line the thoracic cavity and invest the surfaces of the lungs

8. _____

9. the region of the trunk between the diaphragm and the neck

9. _____

10. continuous layer of tissue which lines the pulmonary alveoli

10. _____

11. a fold formed by a sleeve of pleura enclosing the bronchi and pulmonary blood vessels

11. _____

12. closed membranous sac surrounding the heart

12. _____

13. portions of the thoracic cavity (excluding the mediastinum) which are not occupied by the lungs

13. _____

14. in its original place

14. _____

FIGURE 2.1 THE LUNGS AND THE BRONCHIAL TREE.

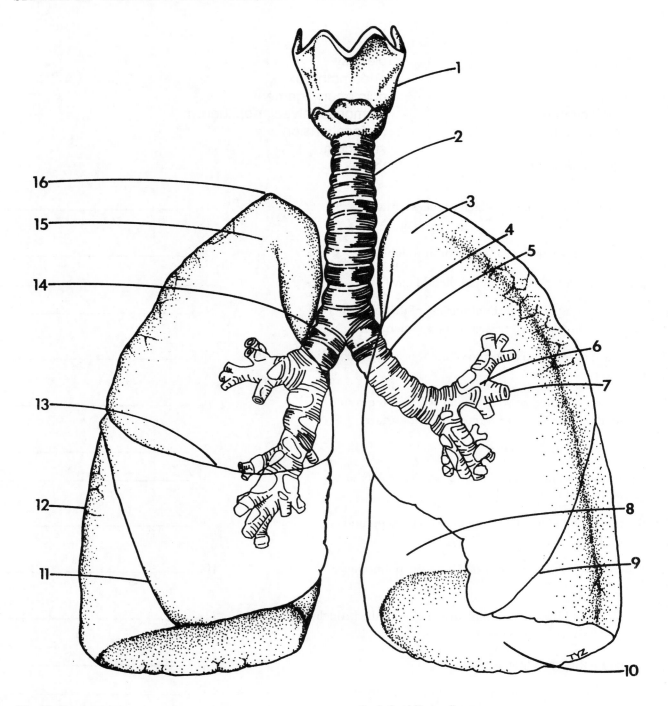

Identify: 1. larynx
2. trachea
3. left lung
4. hilum of left lung
5. left mainstem bronchus
6. secondary bronchus
7. tertiary bronchus
8. cardiac impression

9. left oblique fissure
10. base of lung
11. right oblique fissure
12. costal surface of lung
13. horizontal fissure of right lung
14. right mainstem bronchus
15. right lung
16. apex of right lung

FIGURE 2.2 THORACIC VISCERA

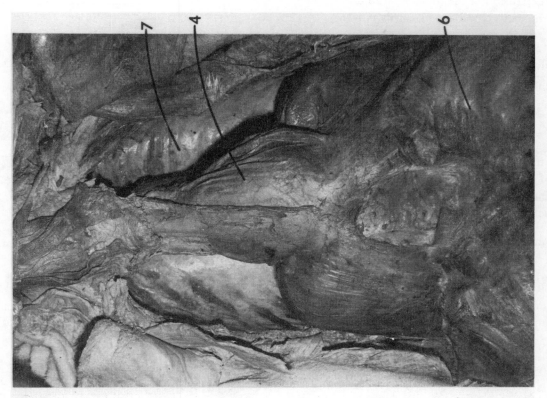

Thoracic cavity with lungs removed.

4. pericardium containing the heart

5. right lung

6. reflected anterior wall of thorax

7. posterior wall of thorax

Thoracic viscera in relation to the diaphragm.

Identify: 1. left lung

2. left dome of diaphragm

3. right dome of diaphragm

No. 2-11 MECHANICAL ASPECTS OF THE PLEURAE
Text pages 42–43

1. With increased expansion of the thoracic wall, the tendency of the lungs to recoil progressively *increases / decreases.*

    1. _____

2. Negative intrapleural fluid pressure links the lungs to the thoracic wall because

    2.

    a. there is, between the pleural membranes, a thin layer of *gas / liquid.*

    a. _____

    b. the pleural membranes are *permeable / impermeable.*

    b. _____

    c. absorption of gases and fluids, particularly through the action of the *visceral / parietal* pleura, generates subatmospheric pressure which links the membranes.

    c. _____

3. Pleural-surface (intrapleural) pressure, which indicates the tendency of the lungs to recoil from the thoracic wall, is measured by means of a wet *spirometer, manometer.*

    3. _____

4. While all gases and most liquids are absorbed by the pleural membranes, a small amount of liquid must remain to _____ the coupling system.

    4. _____

5. Because the growth rate of the lungs *equals / exceeds / is exceeded by* the growth rate of the body as a whole, the stretching forces to which the lungs are subjected *increase / decrease.*

    5. _____
    _____

6. In an adult at resting expiratory level, pleural-surface (intrapleural) pressure is *below, above, equal to* atmospheric pressure.

    6. _____

No. 2-12 THE FRAMEWORK FOR THE BREATHING MECHANISM
Text page 43

1. List the three major components of the skeletal framework for breathing.

    1. _____
    _____
    _____

2. The framework for breathing also includes the skull.

    2. T F

No. 2-13 THE FRAMEWORK FOR THE BREATHING MECHANISM: THE SPINAL COLUMN
Text pages 44-49

Description Vertebrae:	Cervical	Thoracic	Lumbar
axis	✔		
largest			✔
have transverse foramina which form passageway for vertebral artery and vein	✔		
have articular facets for ribs		✔	
atlas	✔		
five in number			✔
twelve in number		✔	
seven in number	✔		
large bodies for weight-bearing			✔
lack features of other vertebrae			✔

1. The fused sacral vertebrae appear as a single bone called the _____.

2. Although the coccygeal vertebrae are not fused, they are referred to as a structure called the _____ or more commonly the _____.

3. The vertebral foramen is occupied by the _____.

4. The spinous process at the base of the neck which is most easily palpated is the 2nd / 5th / 7th cervical vertebra.

5. Movement of the spinal column is restricted by intervertebral _____ and a system of _____.

6. Identify the cervical, thoracic, and lumbar vertebrae.

1. _____

2. _____

3. _____

4. _____

5. _____
   _____

6. a. _____
   b. _____
   c. _____

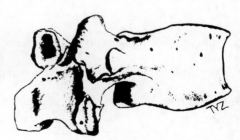

a   b   c

atlas   lordosis   spina bifida
axis   lumbar   thoracic
cervical   odontoid   vertebra
coccygeal   pedicles   vertebral foramen
corpus   rudimentary   vestigial
dens   sacral
kyphosis   scoliosis

1. derived from a word meaning neck

2. incomplete fusion of arches of the vertebral column

3. derived from a word meaning footprint; a remnant of a structure which, in an earlier stage of development or in a preceding organism, was functional

4. derived from a word meaning loin

5. a lateral curvature of the spinal column

6. derived from a word meaning cuckoo

7. derived from a word meaning beginning; an incompletely developed structure

8. second cervical vertebra

9. an upward projection of the body of the axis that provides a pivot around which the atlas and the skull rotate

10. the body of a vertebra

11. first cervical vertebra

12. an abnormally increased convex curvature of the lumbar region

13. the leglike parts of vertebrae

14. hunchback; an abnormally increased concave curvature of the thoracic region

15. derived from a word meaning sacred

16. derived from a word meaning to turn

17. pertaining to chest; region above diaphragm and below the neck

18. a large canal in a vertebra through which the spinal cord passes

1. _____
2. _____
3. _____
4. _____
5. _____
6. _____
7. _____
8. _____
9. _____
10. _____
11. _____
12. _____
13. _____
14. _____
15. _____
16. _____
17. _____
18. _____

FIGURE 2.3 THE SPINAL COLUMN.

Identify:

1. dens of C_2
2. atlas (C_1)
3. axis (C_2)
4. C_7 (seventh cervical)
5. transverse process
6. spinous process
7. sacrum
8. coccyx
9. sacral foramen
10. L_5 (fifth lumbar)
11. T_{12} (twelfth thoracic)

SCHEMATIC OF A VERTEBRAE

Draw a rough schematic of a thoracic vertebrae as seen from above.

Label the following:

corpus (body)

pedicles (legs)

vertebral foramen (canal)

neural arch

spinous process

transverse process

articular facets

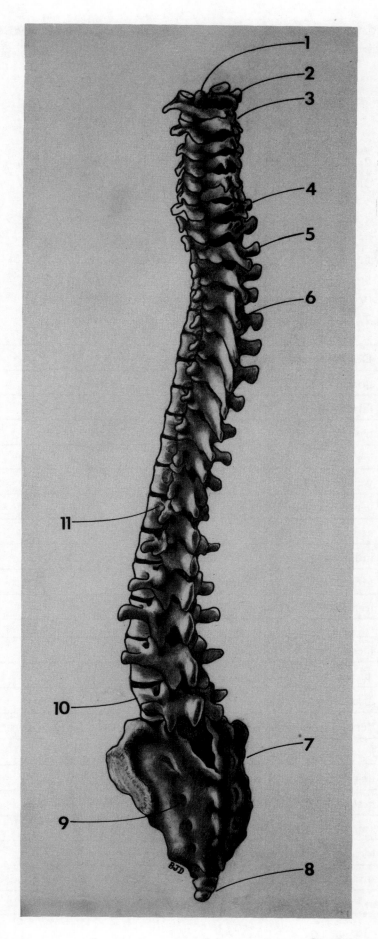

1. The components of the rib cage: 1.
    a. in the midline posteriorly   a. _____
    b. in the midline anteriorly   b. _____
    c. laterally   c. _____

2. The osseous ribs articulate with the *sternum / vertebrae / both*.   2. _____

3. The angle or chondro-osseous union of a rib is generally   3. _____
    superior / inferior to its posterior attachment.

4. In infants the course of the ribs is more *horizontal /*   4. _____
    oblique than in adults.

5. The ribs articulate with the vertebral column by means of   5. _____
    gliding / saddle / pivot joints.

6. Ribs 1, 10, 11, 12 articulate with *one / two* vertebra(e).   6. _____

7. Ribs 2 through 9 articulate with *one / two* vertebra(e).   7. _____

Description of Ribs	1	2–7	8–10	11–12
floating ribs				✔
false ribs			✔	
true ribs	✔	✔		
vertebrochondral ribs			✔	
vertebral ribs				✔
vertebrosternal ribs	✔	✔		
articulate(s) with sternum by synchondrosis	✔			
indirectly connected with sternum by long costal cartilages			✔	
articulate(s) with sternum by synovial joint		✔		
free anterior extremities				✔

Description of Processes of Sternum	Manubrium	Body	Xiphoid
also known as ensiform process			✔
also known as corpus		✔	
derived from word meaning handle	✔		
location of jugular or suprasternal notch	✔		
its uppermost articulation is with clavicle	✔		
first costal cartilage attaches to its lateral border	✔		
second costal cartilage attaches at the junction of	✔	✔	
location of sternal angle indicates junction of	✔	✔	
small cartilaginous process which begins to ossify in adulthood			✔
in early life composed for four sternebrae		✔	

Notes: *Pectus carinatum* (L. keeled breast), also called *pigeon breast,* is characterized by undue prominence of the sternum and the cartilaginous portions of the ribs.

Pectus excavatum (L. hollowed breast), also called *funnel breast,* is characterized by undue depression of the inferior portion of the sternum. Although sometimes caused by rickets or chronic obstruction of respiration, it is usually congenital. In males it may be a manifestation of Fragile X syndrome.

Quote: "Fragile X [fra(X)] syndrome is considered the most common inherited cause of mental retardation in males in the general population, with a calculated prevalence of 1 in 1,000 to 1 in 2,000. Only Down syndrome is seen more commonly but most Down syndrome is not inherited."

Susan H. Black, 1992 (Schopmeyer and Lowe)

FIGURE 2.4 A TYPICAL RIB.

Identify:

1. non-articular part of tubercle
2. articular facet of tubercle
3. neck
4. head
5. costal groove
6. shaft
7. angle

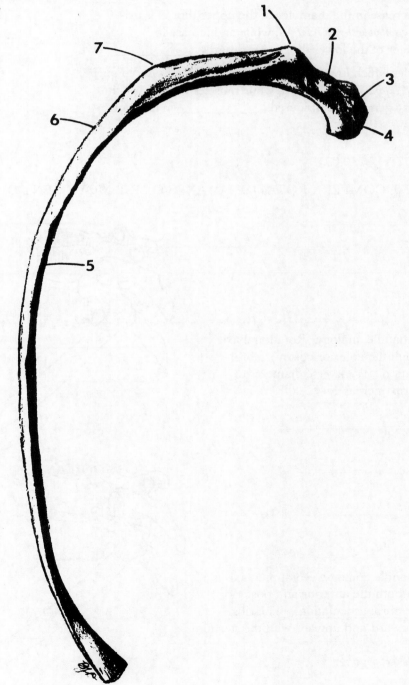

1. The dimensions of the thoracic cavity increase in three planes during *inhalation / exhalation.*

2. The vertical dimension of the thoracic cavity is increased by contraction of the _____.

3. The transverse diameter of the thoracic cavity is increased when the curved ribs are *lowered / raised.*

4. The anteroposterior diameter of the thoracic cavity is increased by simultaneous forward and upward movement of the _____.

5. An increase in the diameter of the upper thorax is primarily *anteroposterior / lateral*, while an increase in diameter of the lower thorax is primarily _____.

6. Muscles which lower the ribs are generally considered *inspiratory / expiratory* in function, while muscles which raise the ribs are generally considered _____ in function.

1. _____

2. _____

3. _____

4. _____

5. _____
   _____

6. _____
   _____

FIGURE 2.5 COMPLEX ROTATIONAL AXES OF RIB MOVEMENT DURING RESPIRATION.

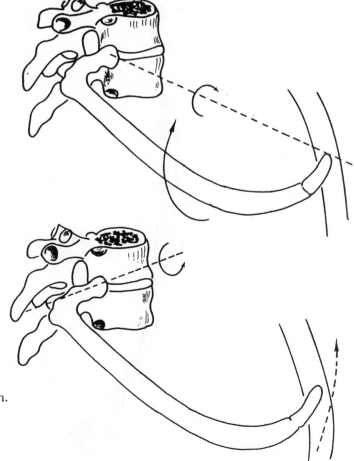

A. Bucket handle analogy. Rotational axis is through the anterior costovertebral articulation. Transverse diameter of the thorax increases.

B. Pump handle analogy. Rotational axis is through both the anterior and posterior costovertebral articulations. The ribs move forward and upward with the sternum.

See Fig 2-34, text page 52.

No. 2-17 THE FRAMEWORK FOR THE BREATHING MECHANISM: THE PELVIC GIRDLE
Text pages 52–54

1. The pelvic girdle, which provides attachment for the lower limbs, is formed by the _____ bones.

2. The bony pelvis is formed by the paired _____ bones plus the _____ and the _____.

3. The socket in the coxal bone which receives the head of the femur is called the _____.

4. The ischial tuberosity is concealed by the _____ muscle.

5. The lower abdomen and the leg are anatomically separated by the _____ ligament.

6. Name the three bones which form each coxal bone.

7. The coxal bone which absorbs body weight during sitting is the _____.

8. The largest coxal bone is the _____.

9. Which coxal bone articulates with the sacrum?

10. How does the pelvis contribute to speech production?

1. _____
2. _____
   _____
   _____
3. _____
4. _____
5. _____
6. _____
   _____
   _____
7. _____
8. _____
9. _____
10. _____

No. 2-18 THE FRAMEWORK FOR THE BREATHING MECHANISM: THE PECTORAL GIRDLE
Text pages 54–55

1. The pectoral girdle is formed by the _____ (collarbone) and the _____ (shoulder blade).

2. The pectoral girdle provides attachment for the _____ to the torso.

3. The large, flat triangular plate of bone which forms the back of the shoulder is called the _____.

4. The strut which projects the scapula clear of the chest wall is called the _____.

5. The free, rather flat projection of the scapula which articulates with the lateral end of the clavicle is called the _____.

6. The space between the medial ends of the clavicles is called the _____ or _____ notch.

1. _____
   _____
2. _____
3. _____
4. _____
5. _____
6. _____
   _____

No. 2-19 RIB CAGE, PELVIC GIRDLE, PECTORAL GIRDLE (VOCABULARY)
Text pages 50–55

acetabulum	inguinal ligament	sternum
acromion	ischium	suprasternal notch
clavicle	jugular notch	vertebral ribs
coracoid	manubrium	vertebrochondral ribs
costal	omo-	vertebrosternal ribs
corpus	pectoral	xiphoid process
coxal bone	scapula	
ensiform process	sternal angle	

1. "swordlike" inferior portion of the sternum

    1. _____

    _____

2. hip bone

    2. _____

3. a palpable projection at the junction of the manubrium and the corpus of the sternum

    3. _____

4. derived from a word meaning like the beak of a crow

    4. _____

5. pertaining to breast

    5. _____

6. pertaining to ribs

    6. _____

7. true ribs

    7. _____

8. floating ribs

    8. _____

9. upper segment of the sternum

    9. _____

10. derived from word meaning hip

    10. _____

11. socket which accepts the thigh bone; "vinegar cup"

    11. _____

12. false ribs

    12. _____

13. collar bone

    13. _____

14. depression on the superior border of the manubrium

    14. _____

    _____

15. breastbone

    15. _____

16. shoulder blade

    16. _____

17. uppermost point of the shoulder

    17. _____

18. meaning body

    18. _____

19. pertaining to shoulder

    19. _____

20. anatomical division between leg and lower abdomen

    20. _____

FIGURE 2.6 THE FRAMEWORK FOR THE BREATHING MECHANISM.

Identify:

1. T_1
2. clavicle
3. spine of scapula
4. vertebrosternal rib
5. crest of ilium
6. ant. sup. iliac spine
7. sacrum
8. coccyx
9. ischium
10. pubic symphysis
11. greater sciatic notch
12. post. inf. iliac spine
13. post. sup. iliac spine
14. L_1
15. vertebral rib
16. angle of rib
17. inferior angle of scapula
18. vertebral border, scapula
19. axillary border, scapula
20. glenoid fossa
21. acromion of scapula
22. C_1

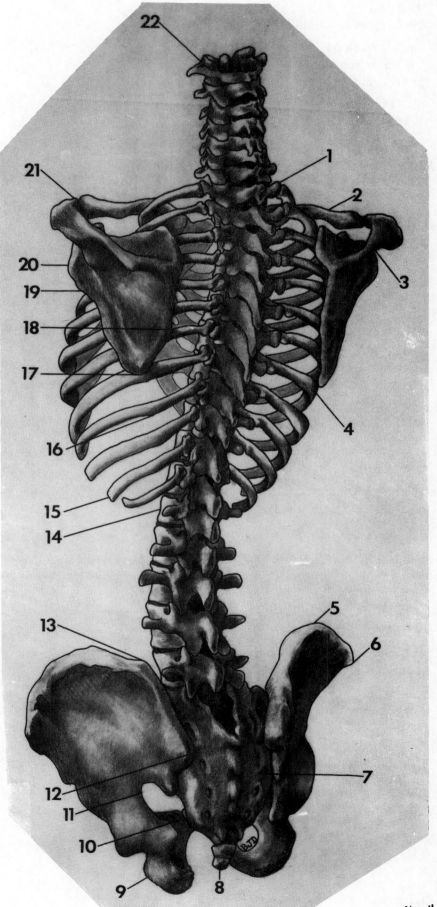

No. 2-20 THE MUSCULATURE OF THE BREATHING MECHANISM: INTRODUCTION
Text page 55

    1. A Cycle of Breathing:

        a. Through the contraction of the thoracic muscles, the size of the thoracic cavity increases in *one / two / three* dimension(s).

        b. The lungs also expand, but by means of _____ linkage.

        c. With thoracic expansion the pressure which is momentarily generated within the pulmonary alveoli is *positive / negative.*

        d. Air rushes into the lungs until the intraalveolar pressure is equal to _____ pressure.

        e. As the muscles of inhalation cease to contract, the dimensions of the expanded thorax-lung complex decrease, thus generating a slightly *positive / negative* intraalveolar pressure, and air is exhaled.

    2. In quiet breathing the forces of expiration are *active / passive* and *muscular / nonmuscular.*

    3. In adults a quiet respiratory cycle occupies approximately *three / five / seven* seconds.

    4. The amount of air exchanged during each cycle of breathing is approximately 500-750 cu. cm. or ____-____ liter(s).

    5. During quiet breathing muscle contraction is required for *inspiration / expiration.*

    6. Forced exhalation is facilitated by contraction of the *abdominal / thoracic* muscles.

1.

  a. _____

  b. _____

  c. _____

  d. _____

  e. _____

2. _____
   _____

3. _____

4. _____

5. _____

6. _____

No. 2-21 THE MUSCLES OF THE THORAX: THE DIAPHRAGM
Text pages 55–58

    1. The diaphragm divides the torso into the _____ and the _____.

    2. The periphery of the diaphragm consists of muscular fibers which originate at the bottom of the rib cage and insert into the edges of an aponeurosis called the _____.

    3. The fibrous pericardium blends into and becomes a part of the _____.

    4. When the diaphragm descends the *lungs / heart / both* also descend(s).

    5. The large organ which is suspended from the diaphragm by five ligaments is the _____.

    6. The diaphragm is a/an *paired / unpaired* muscle.

    7. The diaphragm has a *unilateral / bilateral* nerve supply.

1. _____
   _____

2. _____

3. _____

4. _____

5. _____

6. _____

7. _____

Description Part of the Muscular Diaphragm:	Sternal	Costal	Vertebral
encircles the esophagus			✔
originates from the xiphoid process	✔		
originates from the upper lumbar vertebrae by means of crura			✔
originates from the cartilages of ribs 7–12		✔	
inserts in the central tendon	✔	✔	✔

Structures Passing through Diaphragm Openings:	Aortic Hiatus	Esophageal Hiatus	Foramen Vena Cava
the musculomembranous tube that carries food to the stomach		✔	
the main trunk of the arterial system	✔		
the trunk of the venous system for the lower part of the body			✔

No. 2-22 THE MUSCLES OF THE THORAX: THE INTERCOSTAL MUSCLES
Text pages 58–59

Description Intercostal Muscles:	External	Internal
located between the ribs	✔	✔
eleven on either side	✔	✔
stronger	✔	
more prominent	✔	
muscle fibers continue to the sternum		✔
muscle fibers continue to the vertebral column		
terminate as intercostal membranes near the chondro-osseous unions of the ribs	✔	
terminate as posterior intercostal membranes which continue to the vertebrae		✔
structural variability is an important consideration in EMG studies	✔	
fibers course at right angles to the other intercostals	✔	✔

FIGURE 2.7 THE DIAPHRAGM.

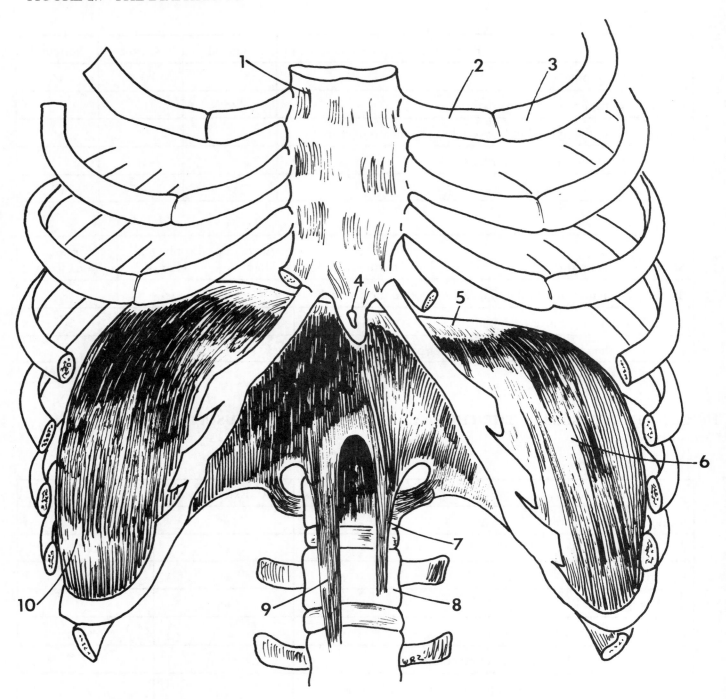

Identify:

1. sternum
2. costal cartilage
3. osseous portion of rib
4. xiphoid process
5. central tendon of diaphragm
6. muscular portion of left hemidiaphragm
7. left crus
8. lumbar vertebra
9. right crus
10. muscular portion of right hemidiaphragm

No. 2-23 MUSCLES OF THE THORAX: SUBCOSTALS, TRANSVERSUS THORACIS, COSTAL ELEVATORS, SERRATUS POSTERIOR
Text pages 59-60

Description Thoracic Muscles:	Subcostals	Transversus Thoracis	Costal Elevators	Serratus Posterior
intracostals	✔			
originate from vertebrae			✔	✔
originate from sternum and fan out		✔		
inferior and superior				✔
breves and longus			✔	
form musculomembranous sheet lining the back of the thorax	✔			
line the inner surface of the anterior thoracic wall		✔		
triangularis sterni		✔		
similar to intercostals, but not confined to one intercostal space	✔			
levatores costalis			✔	
appear to be a continuation of the external intercostals			✔	
often largely aponeurotic or poorly developed; sometimes missing				✔

FIGURE 2.8 MUSCULATURE ON THE VENTRAL SURFACE OF THE TORSO.

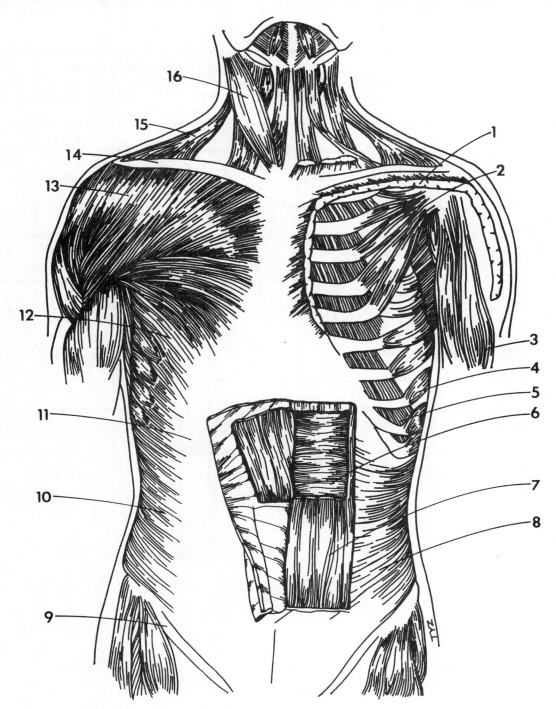

Identify:

1. pectoralis major m.

2. pectoralis minor m.

3. biceps brachii m.

4. serratus anterior m.

5. internal intercostal m.

6. transversus abdominis m.

7. rectus abdominis m.

8. internal oblique m.

9. inguinal ligament

10. external oblique m.

11. abdominal aponeurosis

12. pectoralis major m. (sternal part)

13. pectoralis major m. (clavicular part)

14. clavicle

15. trapezius m.

16. sternocleidomastoid m.

FIGURE 2.9 MUSCULATURE ON THE DORSAL SURFACE OF THE TORSO.

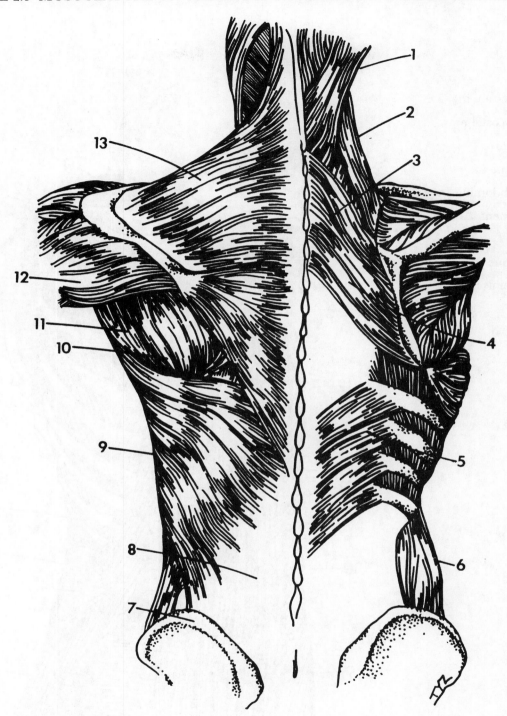

Identify:

1. splenius capitus and cervicus m.
2. levator scapulae m.
3. rhomboideus minor m.
4. rhomboideus major m.
5. serratus post. inf. m.

6. quadratus lumborum m.
7. crest of ilium
8. thoracolumbar fascia
9. latissimus dorsi m.
10. infraspinatus m.

11. teres*
12. deltoideus m.
13. trapezius m.

teres (L. long and round), not to be confused with teras (L. monster). Teratology is the study of birth defects.

FIGURE 2.10 BREATHING MUSCULATURE, ANTEROLATERAL SURFACES OF THE TORSO.

Identify:

1. C$_7$
2. clavicle
3. internal intercostal m.
4. external intercostal m.
5. linea semilunaris
6. internal oblique m.
7. transversus abdominis m.
8. inguinal ligament
9. linea alba
10. external oblique m.
11. rectus abdominis m.
12. fifth rib

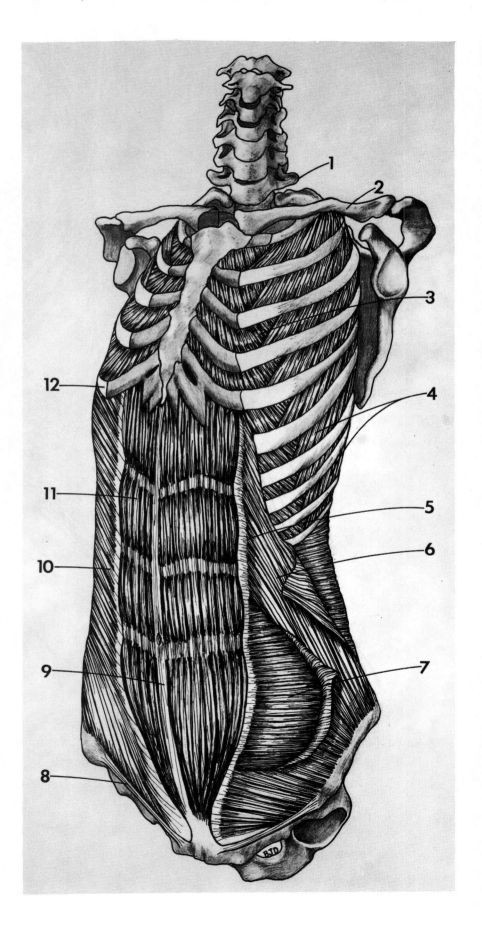

FIGURE 2.11 BREATHING MUSCULATURE, AND SOME DEEP MUSCLES OF THE BACK.

Identify:

1. rhomboideus minor m.
2. rhomboideus major m.
3. levator costalis m., pars brevis
4. serratus posterior inferior
5. sacrum
6. coccyx
7. coxal bone
8. levator costalis m., pars longus
9. external intercostal m.
10. acromion of scapula
11. clavicle
12. T₁

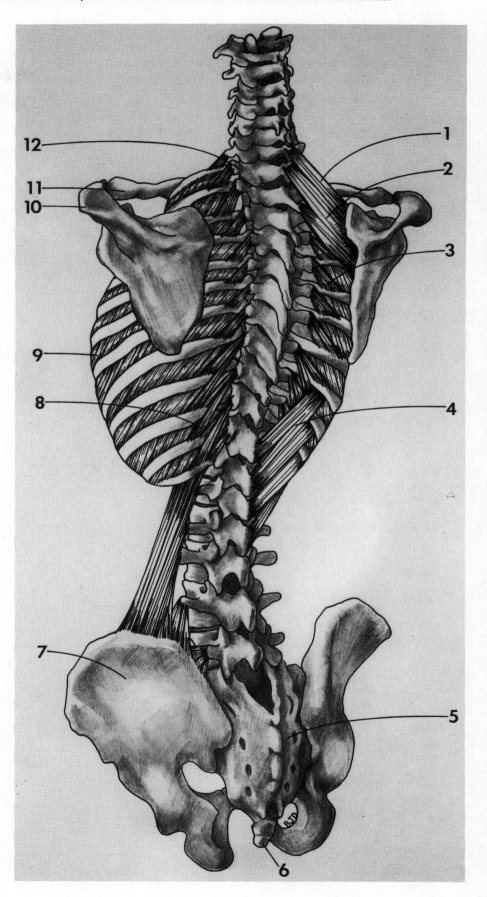

aponeurosis   foramen   levator costalis
breves   forced exhalation   palpation
central tendon   hiatus   pericardium
crura   intercostal   potential energy
diaphragm   kinetic energy   radiography
EMG

1. a broad sheet of tendinous tissue forming the attachment between a muscle and the part that it moves

    1. _____

2. energy of position

    2. _____

3. membranous sac enclosing the heart

    3. _____

4. leglike part

    4. _____

5. an opening

    5. _____

    _____

6. derived from a word meaning partition

    6. _____

7. aponeurotic portion of diaphragm

    7. _____

8. X-ray

    8. _____

9. rib raiser

    9. _____

10. between ribs

    10. _____

11. examination by feeling

    11. _____

12. contraction of abdominal musculature to exhale air beyond that exhaled passively

    12. _____

13. recording changes in electric potential of muscle

    13. _____

14. short

    14. _____

15. energy of motion

    15. _____

Note: *Auscultation* is examination by listening to sounds emanating from various organs, particularly the lungs and the heart.

1. When the diaphragm contracts:

    a. the central tendon is pulled downward and *forward / backward*.

    b. the thorax enlarges primarily *laterally, anteriorly-posteriorly, vertically*.

2. When the diaphragm contracts:

    a. thoracic volume

    b. pressure within the thorax

    c. volume of the abdominal cavity

    d. pressure within the abdominal cavity

3. Vertical excursion of the diaphragm amounts to approximately 1.5 cm or *½ / 1 / 1½* inch(es) during quiet breathing and 6–7 cm or *3 / 4 / 5* inches during deep breathing.

4. The diaphragm contracts throughout the *inhalation / exhalation* phase and momentarily into the *inhalation / exhalation* phase.

5. In some cases the diaphragm may be the only muscle which is really active during quiet breathing.

6. Diaphragmatic (abdominal) breathing requires voluntary control of the diaphragm.

7. The diaphragm may contract toward the end of maximum exhalation.

8. A functional diaphragm is essential for breathing.

9. When does the diaphragm contract very strongly? _____

_____

_____

1.

    a. _____

    b. _____

2.   increases   decreases

    a. _____

    b. _____

    c. _____

    d. _____

3. _____
    _____

4. _____
    _____

5. T   F

6. T   F

7. T   F

8. T   F

Type of Breathing: Description	Abdominal (diaphragmatic)	Costal (thoracic)
base of the thorax expands and abdominal viscera fill up the space created in the abdomen		✔
contraction of the diaphragm compresses the abdominal viscera and results in expansion of the abdominal wall	✔	
becomes more prominent with progressively deeper breathing	✔	
used by most people	✔	✔

No. 2-26 THE ACTION OF MUSCLES OF THE THORAX: THE INTERCOSTAL MUSCLES
Text pages 62–64

1. As a group, the intercostal muscles

   a. contribute to the rigidity of the _____ wall.

   b. help control the degree of space between the _____.

   c. couple the _____, one to another.

   d. function in the flexion of the trunk, thus being classified as _____ muscles.

   e. appear to contribute to sequential control of the breath stream during _____.

2. Characteristics of Class III levers:

   a. always operate with a mechanical *advantage / disadvantage*

   b. power is *lost / gained*

   c. speed is *lost / gained*

   d. *common / uncommon* in the body

3. Each external and internal intercostal muscle forms a Class III lever with an upper rib and with a lower rib. For the external intercostals and the intercartilaginous portion of the internal intercostals, the more efficient lever system is that of the *upper / lower* rib. Contraction of these muscles will thus tend to *elevate / depress* the ribs, thereby aiding *inhalation / exhalation.*

4. It appears that for the interosseous portion of the internal intercostals, the more efficient lever system is that of the *upper / lower* rib. Contraction of these muscles probably _____ the ribs, thus aiding _____.

5. If the external intercostals occupied the intercostal spaces in the chondral portion of the ribs, the tendency of the external intercostals to elevate the ribs would be *increased / decreased.*

6. If the intercostal muscles as a group contribute to the rigidity of the thoracic wall, they will facilitate *inhalation / exhalation / both.*

7. Checking action, which counteracts the elastic recoil of the inflated thorax, consists of prolonged activity of the muscles of inhalation into the _____ phase of breathing.

8. During speech production, as the volume of air in the lungs decreases, the relaxation pressure progressively *increases / decreases.*

9. Checking action continues as long as relaxation pressure *is / is not* providing adequate pressure for speech.

10. When checking action ceases, *inspiratory / expiratory* muscles begin to contract.

1.
   a. _____
   b. _____
   c. _____
   d. _____
   e. _____

2.
   a. _____
   b. _____
   c. _____
   d. _____

3. _____
   _____
   _____

4. _____
   _____
   _____

5. _____

6. _____

7. _____

8. _____

9. _____

10. _____

62 Chapter 2

11. Using electromyography to measure the electrical energy generated by a muscle, a researcher can accurately determine the function of that muscle.

11. T   F

Explain your answer. _____

_____

_____

Quote: "The preponderance of available evidence suggests that the internal intercostal muscles are one of the principal generators for pulsatile variations of typical magnitude. These small and fast acting muscular generators appear to provide bursts of electrical activity in association with their discrete and brief contractions when the conditions of utterance call for rapid and small pressure variations of respiratory origin. Assuming that these bursts cause displacements of the thorax, the available evidence suggests that the muscles of greatest importance to everyday speech are the internal intercostals."

Thomas J. Hixon, 1973 (Minifie, Hixon, and Williams)

No. 2-27 THE ACTION OF MUSCLES OF THE THORAX: SUMMARY
Text pages 60–65

diaphragm	• most important muscle of breathing • contracts throughout inhalation • increases vertical diameter of thorax • contracts strongly during expulsive efforts
external intercostals and intercartilaginous internal intercostals	• elevate ribs • major contributors to inspiration • active during forced exhalation • important role in checking action
interosseous internal intercostals	• may depress ribs and decrease intercostal space • probably active during speech production, particularly on low expiratory reserve air
subcostals and transversus thoracis	• may depress ribs, thus contributing to exhalation
costal elevators	• may have an important role in the elevation of ribs during inhalation • postural
serratus posteriour superior	• may actively elevate ribs or may complement action of other muscles
serratus posterior inferior	• may exert a downward force on the lower ribs during forced exhalation

Note: Pronounced use of muscles of the neck and upper thorax may occur during very deep breathing, or when the lungs or muscles of respiration are not functioning properly.

1. The Sternocleidomastoid Muscle

    a. As its name indicates, this muscle originates on the sternum and clavicle (*cleido-*) and inserts on the mastoid process of the _____ bone.

    b. Because the sternocleidomastoid muscle inserts behind the rotational axis of the head, unilateral contraction will rotate the head toward the *same / opposite* side, and bilateral contraction will *flex / extend* the neck.

    c. When the head is held in a fixed position, bilateral contraction will elevate the _____ and the _____, thus increasing the antero-posterior dimension of the upper thorax and aiding *inhalation / exhalation*.

2. The Scalene Muscles (lateral vertebral muscles)

    a. The deep muscles of the anterolateral region of the neck are divided into two groups, with the scalenes constituting the *inner / outer* groups.

    b. The scalenes course from the *cervical / thoracic* vertebrae to the upper two ribs.

    c. Unilateral contraction will bend the cervical column toward the *same / opposite* side and bilateral contraction will *flex / extend* the cervical column.

    d. With the cervical column fixed, contraction of the scalenes will *elevate / depress* the upper two ribs, thus aiding *inhalation / exhalation*.

1.

    a. _____

    b. _____
       _____

    c. _____
       _____
       _____

2.

    a. _____

    b. _____

    c. _____

    d. _____
       _____

No. 2-29 MUSCULATURE OF THE BREATHING MECHANISM: ANTEROLATERAL AND POSTERIOR ABDOMINAL MUSCLES
Text pages 71–76

Description/Muscle	Abdominal Muscles:	Anterolateral	Posterior
rectus abdominis		✔	
transversus abdominis		✔	
quadratus lumborum			✔
external and internal obliques		✔	
iliacus, psoas major and minor, pyramidalis*			✔
form a wall between the pelvis and the lower margin of the rib cage		✔	
attach to other muscles and the skeleton by means of abdominal aponeurosis and thoracolumbar fascia		✔	

*These muscles are sometimes considered muscles of the lower limb.

No. 2-30 THE ABDOMINAL MUSCULATURE (VOCABULARY)
Text page 70

abdomen   inguinal ligament   linea semilunaris
abdominal aponeurosis linea alba   lumbodorsal fascia

1. a dense, fibrous band extending uninterrupted, except for the umbilicus, from the xiphoid process of the sternum to the pubic symphisis

 1. _____

2. part of the body bounded above by the diaphragm and below by the inlet to the pelvis

 2. _____

3. a broad, two-layered sheet on the dorsal aspect of the lower vertebral column

 3. _____

4. a tendinous structure which defines the separation between the abdominal wall and the lower limb

 4. _____

5. a broad, flat sheet of tendinous tissue on the anterior abdominal wall; extends from sternum to pubis

 5. _____

6. vertical, fibrous bands lateral to the rectus abdominis

 6. _____

Quote: "During the (embryonic) morphogenesis of the muscular system there may be degeneration of portions of (or an entire) muscle segment. Muscles which degenerate in this manner tend to become converted into connec-tive tissue. Many of the large and strong aponeurotic sheets found in the body are formed by this process."

B. M. Patten, 1953

No. 2-31 THE ANTEROLATERAL ABDOMINAL MUSCLES
Text pages 71-75

Description Anterolateral Abdominals:	External Obliques	Internal Obliques	Transversus Abdominis	Rectus Abdominis
deepest			✔	
superficial	✔			✔
largest and strongest	✔			
middle layer of abdominal musculature		✔		
parallel to midline, just lateral to linea alba				✔
postural	✔	✔		✔
compress abdominal contents	✔	✔	✔	✔
probably active in forced exhalation	✔	✔	✔	✔
most suited for electromyographic study	✔			✔
probably the most effective rib depressors		✔		
may most effectively compress abdominal contents			✔	
contraction may limit depth of inspiration	✔	✔	✔	✔

No. 2-32 THE ABDOMINAL MUSCULATURE: THE QUADRATUS LUMBORUM AND ITS ACTION
Text pages 75–76

1. The quadratus lumborum is an *anterolateral / posterior* abdominal muscle.

2. Because of its costal attachments it may be regarded as an active muscle of *inhalation / exhalation*.

3. It may help anchor the lower two ribs against the lifting force of the diaphragm, thus acting as an accessory to *inhalation / exhalation*.

4. It is a flexor of the _____ vertebrae.

1. _____

2. _____

3. _____

4. _____

No. 2-33 THE MECHANICS OF BREATHING: A CYCLE OF QUIET BREATHING
Text pages 76–96

ACTIVE INHALATION

ACTION	EFFECT

• muscle contraction:
    diaphragm
    intercostals
    maybe scalenes
	• increases dimensions of the thorax in all planes

• lungs expand, closely following the enlarging thorax	• air rushes in to equalize pressure
• simultaneously the abdominal viscera are compressed by the descending diaphragm	• intraabdominal pressure is elevated
• muscles of inhalation gradually cease their activity	• restoring forces come into play

PASSIVE EXHALATION

ACTION	EFFECT
• upward force against the diaphragm increases	• vertical dimension of the thorax decreases
• the ribs, which have been elevated, unwind	• provides rotational restoration force (torque)
• system acted upon by gravity	• potential energy will be recovered in the form of kinetic energy
• highly elastic lungs which are linked to the thoracic wall exert a progressive restoring force with increased stretch	• restores thorax to undilated position
• simultaneously lung elasticity provides expiratory force	• air is expelled from lungs

Note: Alveolar pressure is the same as atmospheric pressure at the:

1. beginning of inspiration.
2. end of inspiration.
3. end of expiration.

LUNG VOLUMES

Tidal Volume (TV)	the volume of air inhaled and exhaled during any single respiratory cycle
Inspiratory Reserve Volume (IRV)	the quantity of air which can be inhaled beyond that inhaled in a tidal volume cycle
Expiratory Reserve Volume (ERV) or Resting Lung Volume (RLV)	the amount of air that can be forcibly exhaled following a quiet or passive exhalation
Residual Volume (RV)	the quantity of air that remains in the lungs and airways even after a maximum exhalation

Note: Lung *volumes* are discrete values; no one volume includes another volume.
Lung *capacities* all include two or more of the above volumes.

Resting expiratory level refers to a state of equilibrium in the respiratory system.
The forces of compression of the lungs are balanced by the forces of expansion
of the thorax.

LUNG CAPACITIES

Inspiratory Capacity (IC) = tidal volume + inspiratory reserve volume	the maximum volume of air that can be inhaled from the resting expiratory level
Vital Capacity (VC) = inspiratory reserve volume + tidal volume + expiratory reserve volume	the quantity of air that can be exhaled after as deep an inhalation as possible
Functional Residual Capacity (FRC) = expiratory reserve volume + residual volume	the quantity of air in the lungs and airways at the resting expiratory level
Total Lung Capacity (TLC) = sum of all lung volumes	the quantity of air the lungs are capable of holding at the height of maximum inhalation

AIR EXCHANGE MEASUREMENTS

Minute Volume	liters of air exchanged per minute during quiet breathing (active inhalation, passive exhalation)
Maximum Minute Volume or Maximum Breathing Capacity	the liters of air which would be exchanged if a person could forcefully inhale and exhale for a full minute (usually based on an eight or ten second sample)

No. 2-35 THE MECHANICS OF BREATHING (VOCABULARY)
Text pages 76–81

breath group

residual air

dead air

spirogram

dead air space

spirometer

pulmonary compliance

torque

pulmonary subdivisions

yawn

1. lung volumes and capacities

1. _____

2. syllables produced during single expiratory movement

2. _____

3. an automatic and involuntary deep inhalation which provides needed oxygen to the blood stream

3. _____

4. a graphic recording of lung volume and capacity

4. _____

5. the air, the first to be inhaled and the last to be exhaled, which does not contribute oxygen to the blood or receive carbon dioxide from it

5. _____

6. an instrument for measuring lung volume and capacity

6. _____

7. a rotational restoring force

7. _____

8. the air that cannot be forcibly exhaled and which remains in the lungs after death

8. _____

9. portion of the respiratory tract from the oral and nasal cavities to the bronchioles

9. _____

10. determines degree to which lung/thorax complex can be distended

10. _____

No. 2-36 SPIROGRAMS
Text pages 77–81

If your laboratory is equipped with a wet spirometer or better yet, a computerized measurement system determine the following:
1. tidal volume
2. inspiratory reserve volume
3. expiratory reserve volume
4. vital capacity

Use the values you obtain to complete a spirogram on the next page. Assume a residual volume of 1000 ml.
1. Does the sum of your expiratory reserve, tidal and inspiratory reserve volumes equal your vital capacity?
2. Again, assuming a residual volume of 1000 ml, what is your total lung capacity? functional residual capacity?

Think about the possible characteristics of spirograms for:
1. a person with excessive lung compliance, as in the case of emphysema.
2. a person who has lessened lung compliance, as in silicosis.

What changes would occur in the four pulmonary measures listed above if your entire body, except for your head, were placed in:
1. a pressurized environment?
2. a partial vacuum?

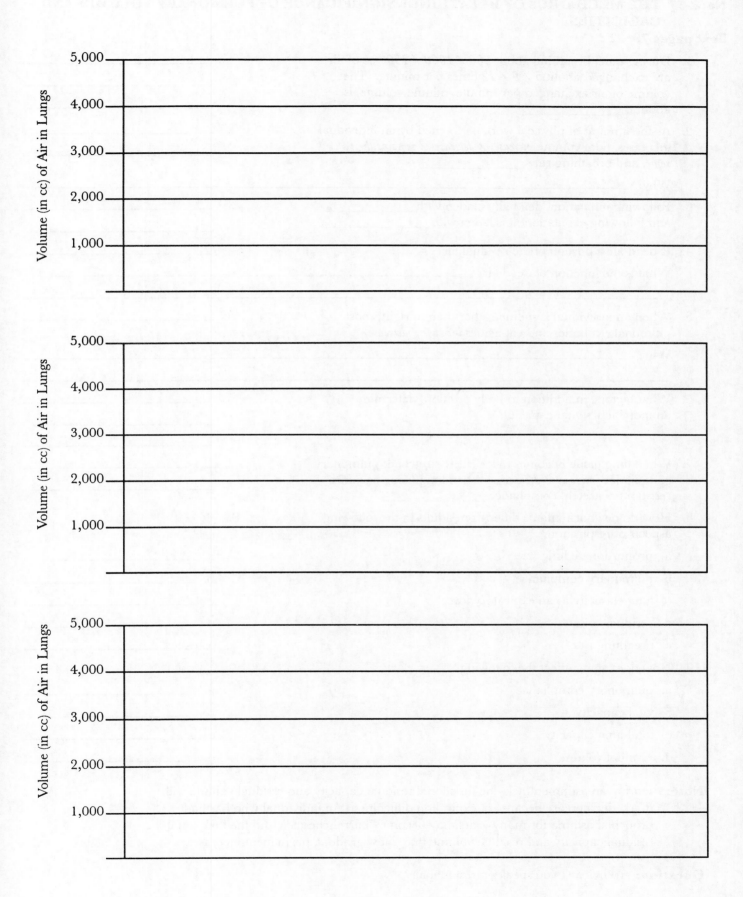

No. 2-37 THE MECHANICS OF BREATHING: SIGNIFICANCE OF PULMONARY VOLUMES AND CAPACITIES
Text pages 79–82

1. During quiet breathing about *500 / 1000 / 1500* cc of air are exchanged about *5 / 9 / 12* times per minute. The volume of air exchanged per minute (minute volume) is about 6 _____.

   1. _____
      _____
      _____
      _____

2. As the amount of physical work performed by an individual increases, tidal volume *increases / decreases / remains the same*, and breathing rate _____.

   Why? _____

   2. _____
      _____

3. Following inhalation, dead air is laden with _____ and following exhalation, it is laden with _____.

   3. _____
      _____

4. Is dead air a portion of the residual air?

   What is the function of dead air? _____
   _____

   4. _____

5. When an individual is supine rather than upright, most pulmonary volumes and capacities *increase / decrease*.

   Why? _____
   _____

   5. _____

6. What factors, in addition to body position, determine an individual's vital capacity?

   6. _____
      _____
      _____

7. When lung tissue becomes more easily stretched, pulmonary compliance *increases / decreases*. When lung tissue becomes resistant to stretch, compliance _____.

   7. _____
      _____

8. How do newborn infants differ from adults in the following aspects of respiration?

   8.

   a. proportionate lung size

   a. _____

   b. pulmonary compliance

   b. _____

   c. lung elasticity as an expiratory force

   c. _____

   d. residual volume

   d. _____

   e. breathing rate

   e. _____

9. How does aging affect the following?

   9.

   a. pulmonary compliance

   a. _____

   b. vital capacity

   b. _____

   c. total lung capacity

   c. _____

   d. residual volume

   d. _____

Note: During an asthma attack, the functional residual capacity and residual volume will markedly increase because of expiration difficulties. An individual who has had bronchial asthma for many years may become "barrel-chested" and the functional residual capacity and residual volume may show gradual, permanent increases.

Question: Why can't you speak on residual air?

Schematic Representation of Lung-Thorax Unit in Equilibrium (Resting Expiratory Level)

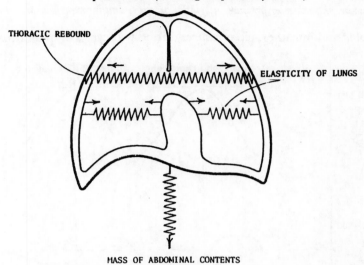

1. Thoracic rebound exerts _____ force on the lungs.

   1. _____

2. Elasticity of the lungs exerts _____ force on the thorax.

   2. _____

3. The mass of abdominal contents exerts _____ force on the diaphragm, and thus on the lungs.
   Why are the lungs subjected to the same gravitational forces as the diaphragm? _____

   3. _____

   _____

4. When the lung-thorax-abdominal system, the functional unit for respiration, is in a state of equilibrium, trans-diaphragmatic pressure is *–1 / 0 / +1.*

   4. _____

5. From the standpoint of mechanics, the cavity which may be regarded as a fluid-filled container is the *thoracic / abdominal* cavity.

   5. _____

6. When the muscles of inhalation expand the thorax the lung 'springs' are increasingly stretched. When thoracic expansion ceases the lung 'springs' will rebound, thus accounting for passive *inhalation / exhalation.*

   6. _____

7. Thoracic 'springs' are compressed by the muscles of forced exhalation. When compression ceases the thoracic 'springs' will rebound, thus contributing to passive *inhalation /exhalation.*

   7. _____

8. Predict what would happen to the following structures if pleural linkage were to hypothetically vanish.

   8.

   a. thoracic walls

   a. _____

   b. lungs

   b. _____

   c. diaphragm

   c. _____

   d. abdominal wall

   d. _____

Note: There are many terms which refer to pressure within the lungs and airways. *Pulmonic pressure, intrapulmonic pressure, alveolar pressure, intraalveolar pressure, subglottal pressure,* and *intratracheal pressure are essentially the same.*

Alveolar and intrathoracic pressures during two cycles of quiet breathing.

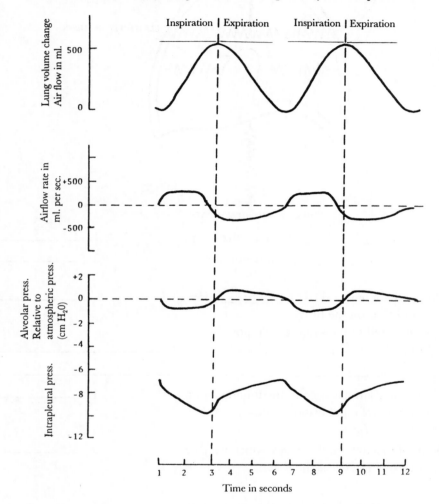

1. Alveolar pressure is the same as atmospheric pressure at the beginning and end of _____, and at the end of _____.

   1. _____
      _____

2. Alveolar pressure is above atmospheric pressure during _____ and below atmospheric pressure during _____.

   2. _____
      _____

3. Intrapleural (pleural-surface) pressure falls during _____ and rises during _____.

   3. _____
      _____

4. The inspiratory phase is usually somewhat *shorter / longer* than the expiratory phase.

   4. _____

5. The slight time lag between changes in alveolar pressure and air flow is due to _____.

   5. _____

1. What are the three mechanisms that alone or in combination regulate alveolar pressure?

    1. _____
    _____
    _____

2. Pressure which is generated entirely by passive forces is called _____ pressure.

    2. _____

3. At resting expiratory level, relaxation pressure is *positive / negative / zero*.

    3. _____

4. During quiet tidal breathing, relaxation following inhalation produces *positive / negative* relaxation pressure which provides the force for *active / passive* exhalation. The forces of relaxation exert pressure until alveolar pressure is equal to _____ pressure.

    4. _____
    _____
    _____

5. Although relaxation pressure is generally considered an expiratory force, it may provide the force for passive inhalation at very *low / high* lung volumes.

    5. _____

6. The relaxation-pressure curve graphically depicts the relationship between _____ volume and _____ pressure.

    6. _____
    _____

7. The relationship between lung volume and relaxation pressure is *linear / nonlinear* at midvolume and *linear / nonlinear* at the extremes of lung volume. At the extremes of lung volume, pressure changes more *slowly / abruptly*.

    7. _____
    _____
    _____
    _____

8. The relaxation pressure curve can be resolved into relaxation pressures generated by the _____ and by the _____.

    8. _____
    _____

9. The rapid changes in relaxation pressure at extremely low lung volumes can be attributed to the *lung / chest wall*, and at extremely high lung volumes to the _____.

    9. _____
    _____

10. At midvolume the lungs and chest wall contribute about equally to changes in relaxation pressure.

    10. T   F

11. At lung volumes above 38% vital capacity
    a. inspiration is
    b. expiration is

    11. Active   Passive
    _____
    _____

12. At lung volumes below 38% vital capacity
    a. inspiration is
    b. expiration is

    12.
    _____
    _____

13. The maximum expiratory and inspiratory pressures that can be generated at any given lung volume represent the algebraic sum of the _____ and _____ forces.

    13. _____
    _____

14. When maximum expiratory and inspiratory pressures are graphed along with the relaxation-pressure curve the entire figure is called the _____ diagram.

    14. _____

15. Alveolar pressure can vary from −150 cm H_2O during a _____ to +200 cm H20 during a _____.

15. _____

_____

Pressure-volume diagram of breathing.

Inspiratory pressures (I_p), relaxation pressures (R_p) and expiratory pressures (E_p) are based on data of Rahn et al., 1946

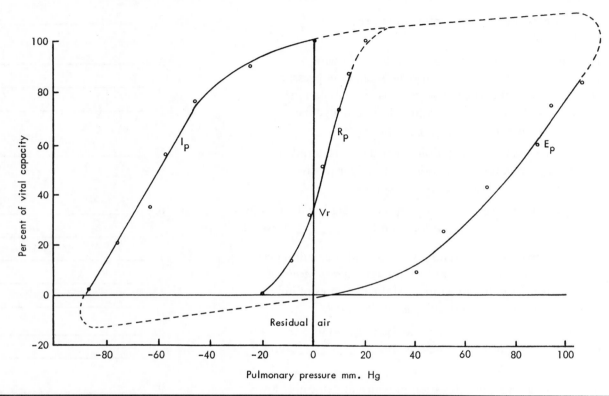

Expiratory Pressure	Inspiratory Pressure	Relaxation Pressure	Vital Capacity			
			0%	38%	80%	100%
+65 mm/Hg	−65 mm/Hg	0		✔		
maximum	0	+30 mm/Hg				✔
0	maximum	−20 mm/Hg	✔			
+100 mm/Hg	−40 mm/Hg	+18 mm/Hg			✔	

16. ". . . pressure equal to the relaxation pressure at a given lung volume cannot be used as evidence for a lack of muscular effort." Hixon, 1973

Why? _____

_____

_____

No. 2-41 THE MECHANICS OF BREATHING: THE EFFECTS OF AIR FLOW RESISTANCE
Text pages 88–90

In the following exercise specify the degree of air flow resistance, low, substantial, or absolute. If it is substantial or absolute list the site or sites of resistance e.g., vibrating vocal folds, compressed lips.

Air-Flow Resistance	Degree	Site(s) of resistance
1. during quiet tidal breathing	_____	_____
2. during thoracic fixation	_____	_____
3. during forced exhalation	_____	_____
4. during forced inhalation	_____	_____
5. below level of vocal folds	_____	_____
6. in nasal cavity when soft palate is depressed	_____	_____
7. production of neutral vowel	_____	_____
8. production of [b] as in boy	_____	_____
9. production of [p] as in pie	_____	_____
10. production of [s] as in see	_____	_____
11. production of [z] as in zoo	_____	_____

Ohm's law

$$\text{current flow} = \frac{\text{voltage}}{\text{resistance}}$$

Analogy applied to speech mechanism*

$$\text{air flow} = \frac{\text{alveolar pressure}}{\text{airway resistance}}$$

(1) Current flow is directly proportional to _____ and inversely proportional to _____.

(1) _____
    _____

(2) Air flow is directly proportional to _____ and inversely proportional to _____.

(2) _____
    _____

(3) If relatively constant air flow is to be maintained, an increase in airway resistance will require a/an *increase / decrease* in alveolar pressure.

(3) _____
    _____

*cannot be applied mathematically because units are not comparable

Questions: If you had inadequate velopharyngeal closure, why might you substitute glottal attacks and glottal stops for stops and fricatives?

What might be the causes and effects of increased air flow resistance in the nasal passageways?

2-42 THE MECHANICS OF BREATHING: PRESSURE AND AIR FLOW REGULATION DURING SPEECH
Text pages 90–93

1. What techniques can be used to measure subglottal pressure during speech? (two answers)

   1. _____

   _____

2. During production of a sustained vowel, subglottal pressure is *highly variable / relatively constant*.

   2. _____

3. Relaxation pressure is in excess of the demands of speech at *low / high* lung volumes.

   3. _____

4. The muscular activity which prevents thorax-lung recoil from generating excessive subglottal pressure is called _____.

   4. _____

5. As speech intensity increases, subglottal pressure *increases / decreases* as does resistance to air flow at the laryngeal level.

   5. _____

6. As airway resistance increases, the degree of checking action needed to maintain a given subglottal pressure *increases / decreases*, thus demonstrating a/an *direct / inverse* relationship between airway resistance and checking action.

   6. _____

   _____

7. Although some people may breathe more deeply during speech the amount of air required for speech is not markedly greater than the amount of air exchanged during quiet breathing.

   7. T   F

8. Lung volume for speech (usually 35-70% of vital capacity) is within the *linear / nonlinear* portion of the relaxation pressure-curve. The demands on checking action are greater within the _____ portion of the curve.

   8. _____

   _____

9. Are the demands placed on the checking action phenomenon generally greater for speaking or singing?

   9. _____

10. Higher lung volumes are usually required during the initiation of *loud / soft* speech.

    10. _____

No. 2-43 THE MECHANICS OF BREATHING: CHEST WALL PREPARATION FOR SPEECH
Text pages 93–94

1. Chest wall preparation for speech facilitates the generation of rapid _____ changes.

   1. _____

2. During chest wall preparation for speech

   2.

   a. the rib cage expands.

   a. T   F

   b. the abdomen expands.

   b. T   F

   c. lung volume increases substantially.

   c. T   F

Quote: "It is virtually impossible to test the phonatory system in isolation from the respiratory. Assessment of the origin of dysphonias is complicated by the well-known fact that respiratory abnormalities affect phonation and phonatory abnormalities affect respiration. Fortunately, in actual practice the phonatory system is statistically far more often implicated in neurologic disease than the respiratory, and most of the phonatory abnormalities one hears stem from direct involvement of the laryngeal musculature."

Darley, Aronson, and Brown, 1975

Characteristics	Diaphragmatic (abdominal)	Thoracic (costal)	Clavicular	Oppositional
lateral expansion of rib cage		✔		
elevation of rib cage and shoulders			✔	
protrusion of abdominal wall	✔			
apparent simultaneous contraction of muscles of inhalation and exhalation; lack of sequential control of breathing musculature				✔
both considered normal (predominant expansion may vary from person to person)	✔	✔		

1. Some simultaneous contraction of muscles of inhalation and exhalation may be used during speech for control purposes.

1. T F

2. During speech the pulse-like control over subglottal pressure is probably due to action of the *intercostals / diaphragm.*

2. _____

3. The use of forced exhalation to provide adequate breath pressure for speech suggests that lung compliance may be *excessive / insufficient.*

3. _____

4. Shallow breathing and frequent breath pauses during speech may suggest that lung compliance is *excessive / insufficient.*

4. _____

5. The duration of phases of inhalation and exhalation is approximately equal during *speech / quiet breathing / both.*

5. _____

6. Exhalation phases are longer than inhalation phases during *speech / singing / both.*

6. _____

Attempt to demonstrate some of the types of breathing described in the clinical note on page 94.
Do not try hyperpnea; you might hyperventilate.

Quote: "Opposition breathing is often observed in cases of cerebral palsy. Excessive activity of the diaphragm is usually coupled with reduced and asynchronous activity of the thoracic muscles. There is very little expansion of the chest during frequent but relatively short periods of inspiration and expiration. Because of rapid, shallow, and arrhythmic breathing, the intensity of phonation may be unusually low, and speech production tends to be erratic in various respects. For example, more breath groups may be required for a given sentence than would be used by normal speakers, and the sentence may be broken up in a manner unrelated to its grammatical structure, giving the listener the impression of choppy, explosive speech."

John K. Darby Jr., 1980

No. 2-45 RESPIRATORY DISORDERS

An inadequate supply of air to the alveoli may be caused by:

1. greater resistance within the airway
   a. obstruction of the bronchioles as in emphysema
   b. swelling of the bronchioles as in asthma
2. loss of tissue elasticity
   a. emphysema
   b. cystic (pulmonary) fibrosis (disorder of mucus-secreting glands)
3. Lessoned compliance of lungs and thorax
   a. tuberculosis
   b. silicosis (caused by inhalation of irritating dust particles; usually an occupational disease)
   c. pneumonia

Inadequate oxygenation of the blood may be caused by:

1. extremely severe lung disease
2. pulmonary edema (abnormal) accumulation of fluid in the cavities and tissue of the lung)
   Marked reduction of hemoglobin in the blood may result in cyanosis (Gr. blue + process)

Inadequate strength or control of the breathing musculature may be caused by:

1. degenerative diseases which cause muscle weakness (may impair the regulation of subglottal pressure)
   a. muscular dystrophy (Gr. disordered + nourishment)
   b. amyotrophic lateral sclerosis (Gr. a-, negative; myo-, muscle)
   c. myasthenia gravis (Gr. muscle + weakness; severe)
2. neurological disorders affecting the timing and strength of muscle contraction
   a. cerebral palsy
   b. different forms of chorea (Gr. dance), characterized by excess involuntary movement

Respiratory depression or arrest may be caused by:

1. drug overdose
2. concussion and the resultant swelling of the brain
3. blockage of the flow of blood to the respiratory center in the medulla
   a. brain injury
   b. brain tumor
   c. stroke

Alternating periods of deep and shallow breathing may be caused by:

1. brain damage
2. heart failure

Note: Newborn infants, especially those born prematurely, may suffer respiratory distress syndrome, also called *hyaline membrane disease*. Because they are unable to produce enough surfactant, their alveoli collapse after each breath.

Quote: "As speech pathologists, it is imperative that we recognize symptoms that may be related to potentially severe pathological conditions. To treat a patient prior to medical referral is a form of professional insanity. To the speech pathologist, medical referral is the first and in many instances the only treatment for patients with suspected bronchial or pulmonary pathology."

Daniloff, Schuckers, Feth, 1980

Chapter 3
Phonation

No. 3-1 PHONATION: INTRODUCTION
Text page 101

1. The principal structure for producing a vibrating air stream is the _____. Its vibrating elements are the _____.

   1. _____
   _____

2. Location of the larynx:

   2.

   a. immediately superior to the _____

   a. _____

   b. immediately inferior to the _____

   b. _____

   c. in an adult, approximately on a level with cervical vertebrae _____ to _____

   c. _____ to _____

   d. *anterior / posterior* neck region

   d. _____

3. The approximate anterior attachment of the vocal folds is indicated by the _____ notch.

   3. _____

No. 3-2 FUNCTIONS OF THE LARYNX: BIOLOGICAL
Text page 101

1. Biologically, the larynx is part of the _____ system.

   1. _____

2. List three protective functions of the larynx. _____
   _____
   _____

3. Active closure of the vocal folds contributes to _____ fixation.

   3. _____

4. Will a person whose larynx has been surgically removed be able to develop intraabdominal pressures adequate for normal biological functions?

   4. _____

Note: Because water beginning to enter the trachea triggers an extremely powerful *laryngeal reflex* resulting in spastic closure of the vocal folds, some drowning victims have no water in their lungs.

No. 3-3 NONBIOLOGICAL FUNCTIONS OF THE LARYNX: MECHANICS OF THE SOUND GENERATOR
Text pages 101–102

1. The larynx can function as a speech sound generator at the same time it is functioning biologically as a protective mechanism.

1. T   F

2. During normal inhalation and exhalation the space between the vocal folds is relatively *narrow / wide*.

2. _____

3. The tone generated by the vibrating vocal folds is called a _____ tone.

3. _____

4. Describe, in your own words, one cycle of vocal fold vibration.

_____

_____

_____

_____

_____

_____

Without referring to your written description, explain vocal fold vibration to another student.

5. The rate of vocal fold vibration during normal vowel production:

5.

   a. in men, approximately _____ Hz (cycles per second)*

a. _____

   b. in women, approximately _____ Hz (cycles per second)

b. _____

6. In a glottal stop the laryngeal adductor musculature suddenly *contracts / releases*. In a glottal attack the musculature suddenly _____.

6. _____

*The term *hertz*, abbreviated Hz, has generally replaced the term *cycles per second*, abbreviated cps. It is an eponym, used in honor of the nineteenth century physicist, Heinrich Hertz, who discovered radio waves.

No. 3-4 THE SUPPORTIVE FRAMEWORK OF THE LARYNX: THE HYOID BONE
Text pages 102–104

1. The hyoid bone is part of the *axial / appendicular* skeleton.

   1. _____

2. The hyoid bone is a supportive structure for the root of the _____.

   2. _____

3. The hyoid bone is the inferior attachment for most of the _____ muscles.

   3. _____

4. The hyoid bone is the superior attachment for *intrinsic / extrinsic* laryngeal muscles.

   4. _____

5. Why is the hyoid bone unique? _____

6. Muscles which attach to the hyoid bone and suspend it in position are called the hyoid _____ muscles or the _____ muscles of the neck.

   6. _____
      _____

7. The hyoid bone is located at the level of the *first / third / sixth* cervical vertebra.

   7. _____

8. The largest part of the hyoid bone is the body or _____.

   8. _____

9. The posteriorly directed limbs of the hyoid bone are called the _____.

   9. _____

10. The prominences at the junction of the body and limbs of the hyoid bone are called the _____.

    10. _____

No. 3-5 THE CARTILAGINOUS FRAMEWORK OF THE LARYNX
Text pages 104–105

Cartilage	Paired	Unpaired	Small	Large	Hyaline	Elastic
thyroid		✔		✔	✔	
cricoid		✔		✔	✔	
epiglottis		✔		✔		✔
arytenoid	✔		✔		✔	
corniculate	✔		✔			✔
cuneiform	✔		✔			✔

Characteristics of hyaline structures at different age levels:

Early years	Mid-years	Later years
soft, flexible capable of interstitial growth	slowly ossifying	bonelike brittle

aditus	glottal arrest	mylo-
arytenoid	glottal attack	omo-
atavism	glottal chink	pars
colliculus	glottal stop	raphe
commissure	glottal stroke	rima glottidis
cornua	glottis	stylo-
cricoid	glottis spuria	symphysis
cuneiform	hyoid	synergy
deglutition	laryngeal prominence	thyroid
digastric	mandible	thyroid prominence
genio-	mental	triticial

*1. seam, suture

1. _____

*2. a coming together

2. _____

*3. the point of union between two structures; a growing together

3. _____

4. like a grain of wheat

4. _____

5. false glottis

5. _____

6. Adam's apple (2)

6. _____

_____

7. having two bellies

7. _____

8. ringlike

8. _____

9. combined action

9. _____

10. forcible approximation of the vocal folds to arrest vibration (2)

10. _____

_____

11. pertaining to chin (2)

11. _____

_____

12. the variable opening between the vocal folds (3)

12. _____

_____

_____

13. the lower jaw

13. _____

14. a throwback; reappearance of remote ancestral characteristics

14. _____

15. U-shaped

15. _____

16. pertaining to pillar

16. _____

17. derived from words meaning wedge + form

17. _____

*These are very similar, but there is a preferred answer to each one.

No. 3-6 cont'd

18. pertaining to molar; derived from word meaning mill

18. _____

19. shieldlike

19. _____

20. sudden initiation of vocal-fold vibration (2)

20. _____

_____

21. derived from words meaning ladle + form

21. _____

22. a part

22. _____

23. an entrance

23. _____

24. swallowing

24. _____

25. the true vocal folds and the opening between them

25. _____

26. horns

26. _____

27. small eminence

27. _____

28. pertaining to shoulder

28. _____

No. 3-7 THE CARTILAGINOUS FRAMEWORK OF THE LARYNX: THE THYROID CARTILAGE
Text pages 104–105

1. The region where the thyroid laminae join is called the _____ of the thyroid.

1. _____

2. The superior thyroid notch is just above the _____.

2. _____

3. The angle of the thyroid is approximately 80 degrees in the adult male and *90 / 120 / 150* degrees in the adult female.

3. _____

4. The superior thyroid horns are usually *shorter / longer* than the inferior thyroid horns.

4. _____

5. The superior thyroid horns are loosely attached to the greater horns of the _____ bone.

5. _____

6. The inferior thyroid horns articulate with the _____ cartilage.

6. _____

Question: What factors might account for the more prominent Adam's apple in males?

No. 3-8 THE CARTILAGINOUS FRAMEWORK OF THE LARYNX: THE CRICOID CARTILAGE
Text page 105

1. The cricoid cartilage forms the *lower / upper* framework
    of the larynx.

1. _____

2. Inferiorly, the cricoid cartilage attaches to the
    _____ tracheal ring by means of the _____
    membrane or ligament.

2. _____
   _____

3. The cricoid cartilage has been likened to a signet ring. The
    anterior part of the cricoid cartilage resembles the *signet / band*
    of the ring, and is called the _____. The posterior
    part resembles the ring's _____, and is called
    the _____.

3. _____
   _____
   _____
   _____
   _____

4. The cricoid cartilage and the inferior horns of the thyroid
    cartilage articulate by means of diarthrodial *gliding / pivot /
    ball and socket* joints. This permits rotation about an axis by
    the thyroid / the cricoid / either cartilage.

4. _____
   _____

No. 3-9 THE CARTILAGINOUS FRAMEWORK OF THE LARYNX:
**   THE ARYTENOID AND CORNICULATE CARTILAGES**
Text page 106

1. The arytenoid cartilages are tetrahedral in form, or
    somewhat like a _____.

1. _____

2. The medial surfaces of the arytenoid cartilages form part
    of the intercartilaginous border of the _____.

2. _____

3. The muscular process of the arytenoid cartilage articulates
    with the _____ cartilage.

3. _____

4. The vocal ligament inserts on the _____ process of
    the arytenoid cartilage. This process is a projection of
    the cartilage's *anterior / posterior* angle near its *apex / base*.

4. _____
   _____
   _____

5. Important laryngeal musculature attaches to the arytenoid
    cartilage's _____ process.

5. _____

6. The apexes of the arytenoid cartilages are capped by
    the _____ cartilages.

6. _____

Pay particular attention to Fig. 3-7, text page 107, which depicts the relationship of the thyroid, cricoid, and arytenoid cartilages.

FIGURE 3.1 THE SKELETAL FRAMEWORK OF THE LARYNX. THE HYOID BONE AND INDIVIDUAL LARYNGEAL CARTILAGES.

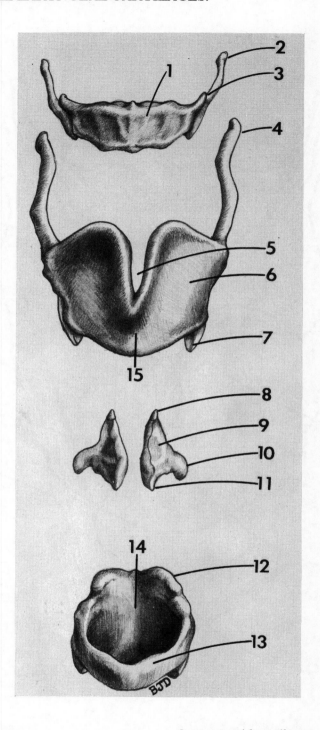

Identify:

1. body (corpus) of hyoid bone
2. greater horn (cornu) of hyoid
3. lesser horn of hyoid bone
4. superior horn of thyroid cartilage
5. thyroid notch
6. left lamina of thyroid cartilage
7. inferior horn of thyroid cartilage
8. corniculate cartilage
9. arytenoid cartilage
10. muscular process of arytenoid cartilage
11. vocal process of arytenoid cartilage
12. cricoarytenoid articular facet
13. arch of cricoid cartilage
14. posterior cricoid lamina (the signet)
15. angle of thyroid cartilage

FIGURE 3.2 THE SKELETAL FRAMEWORK OF THE LARYNX.

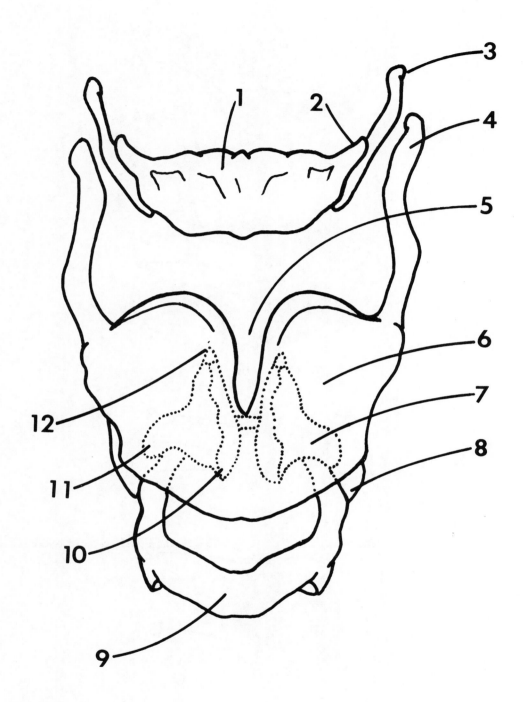

Identify:

1. body (corpus) of hyoid bone
2. lesser horn (cornu) of hyoid bone
3. greater horn of hyoid bone
4. superior horn (cornu) of thyroid cartilage
5. thyroid notch
6. left lamina of thyroid cartilage

7. arytenoid cartilage
8. inferior horn (cornu) of thyroid cartilage
9. arch of cricoid cartilage
10. vocal process of arytenoid cartilage
11. muscular process of arytenoid cartilage
12. corniculate cartilage

No. 3-10 THE CARTILAGINOUS FRAMEWORK OF THE LARYNX: THE EPIGLOTTIS
Text pages 107–110

Note: The prefix *epi-* means on, upon, or over. *Glosso-* refers to the tongue.

1. The epiglottis lies just behind the _____ bone and
    the _____ of the tongue.

    1. _____
    _____

2. The epiglottis attaches to the _____ cartilage and
    the _____ bone.

    2. _____
    _____

3. The epiglottis is more easily observed in *adults / children.*

    3. _____

4. As their name implies, the glossoepiglottic folds (one
    median and two lateral) connect the _____ and the
    _____.

    4. _____
    _____

5. The two pits between the epiglottis and the root of the
    tongue, one on either side of the median glossoepiglottic
    fold, are called _____.

    5. _____

6. The lower half of the anterior (lingual) surface of the
    epiglottis is separated from the hyoid bone, the
    thyrohyoid ligament, and the thyroid cartilage by a
    relatively large _____.

    6. _____

7. Is the epiglottis a vital organ in human beings?

    7. _____

8. Is the epiglottis an important part of the speech mechanism?

    8. _____

9. The epiglottis may prevent food from entering the _____.

    9. _____

10. During speech production, changes in the shape and position
     of the epiglottis seem related to changes in *pitch / intensity.*

    10. _____

Using your hands, attempt to demonstrate the relationship between the vocal folds and the epiglottis.

No. 3-11 THE CARTILAGINOUS FRAMEWORK OF THE LARYNX: THE CUNEIFORM CARTILAGES
Text page 110

1. The folds which extend from the sides of the epiglottis to the
    apexes of the arytenoid cartilages are called the _____
    folds. These folds form the entrance to the _____.

    1. _____
    _____

2. The small, wedge-shaped cartilages embedded within the
    aryepiglottic folds are called the _____ cartilages.

    2. _____

3. The highlighted elevations on the aryepiglottic folds
    which may be observed during the laryngeal examination
    are the _____ tubercles.

    3. _____

4. The cuneiform cartilages, which may be absent in some
    specimens, are *rudimentary / vestigial* structures.

    4. _____

5. The cuneiform cartilages may support and stiffen the
    _____ folds, thus helping maintain the opening
    to the _____.

    5. _____
    _____

FIGURE 3.3 TOP VIEW OF LARYNX, TONGUE AND ASSOCIATED STRUCTURES.

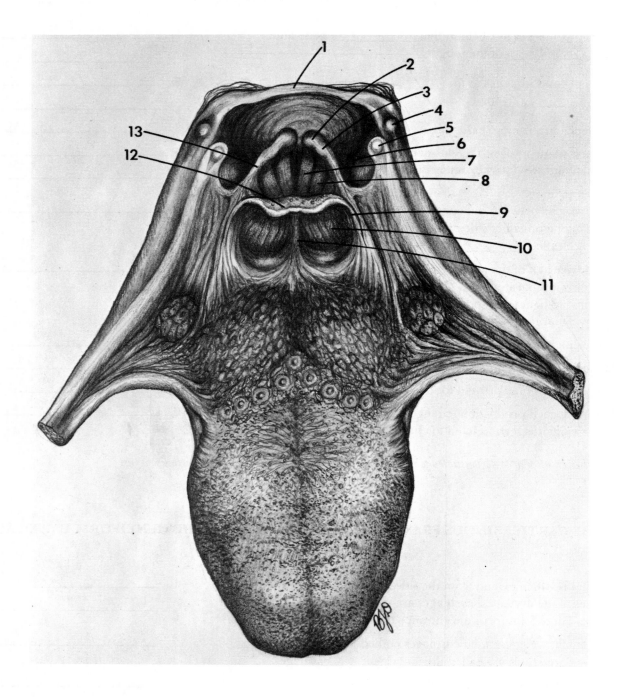

Identify:

1. posterior pharyngeal wall
2. highlight of corniculate cartilage
3. highlight of cuneiform cartilage
4. tip of greater horn of hyoid
5. tip of superior horn of thyroid
6. pyriform sinus
7. true vocal fold
8. false vocal fold
9. lateral glossoepiglottic fold
10. vallecula
11. middle glossoepiglottic fold
12. epiglottis
13. aryepiglottic fold

No. 3-12 THE LARYNGEAL JOINTS: THE CRICOARYTENOID JOINTS
Text pages 110–111

1. The larynx has *two / four* pairs of joints. Why are   1. _____

   they important? _____   _____

2. The cricoid articular facet, located on the superior   2. _____
   border of the cricoid *arch / lamina*, is *convex / concave*.
   _____

3. The arytenoid articular facet, located on the underside   3. _____
   of the *vocal / muscular* process, is *convex / concave*.
   _____

4. The rocking motion of the arytenoid cartilages produces   4. _____
   an upward and outward swinging motion as the vocal
   processes are *abducted / adducted*, and an inward and   _____
   downward swinging motion as the vocal processes are

   _____.

5. The cricoarytenoid joint permits rocking motion and   5. _____
   limited gliding action because it is a/an _____
   joint.

6. Can the rocking and gliding movements of the cricoarytenoid 6. _____
   joint occur simultaneously?

7. The movements of the arytenoid cartilages may be restricted 7. _____
   by the cricoarytenoid _____.

*With your hands representing the arytenoid cartilages and your wrists representing the cricoarytenoid joints,
attempt to demonstrate the movement of the arytenoid cartilages.*

Note: Ankylosis (*ankylo-*, crooked + *-osis*, a process, often pathological), is the fusion
of a joint due to disease. Ankylosis of the cricoarytenoid joint may be caused
by infection, injury, or arthritis, particularly rheumatoid arthritis. If one or both
arytenoid cartilages become fixed the vocal folds will not achieve adequate closure
and the voice may become very weak.

No. 3-13 THE LARYNGEAL JOINTS: THE CRICOTHYROID JOINT
Text pages 112–114 *See also Fig. 3-7 text page 107.*

1. The articular facets of the cricothyroid joints are   1. _____
   located on the cricoid *arch / lamina*, and on the medial
   surfaces of the *superior / inferior* thyroid horns.   _____

2. The predominant motions of the cricothyroid joint   2. _____
   are *rotational / rocking / gliding*. What effect do these
   movements have on the vocal folds?   _____

   _____

3. Movements of the cricothyroid joint are restricted   3. _____
   by the _____ ligaments.

No. 3-14 THE MEMBRANES AND LIGAMENTS OF THE LARYNX
Text pages 114–116

1. The ligaments and membranes connecting the laryngeal cartilages with adjacent structures are called _____ laryngeal membranes.

1. _____

2. The ligaments and membranes interconnecting the laryngeal cartilages are called _____ laryngeal membranes.

2. _____

Extrinsic Laryngeal Membranes

1. The broad membrane seeming to suspend the larynx from the hyoid bone is the _____ membrane. The cordlike elastic ligament, contiguous with the posterolateral border of this membrane, is called the lateral _____ ligament. The small nodule that may be embedded within this ligament is the _____ cartilage.

1. _____
_____
_____

2. As its name implies, the hyoepiglottic ligament, a midline structure, connects the _____ and the _____.

2. _____
_____

3. The lower border of the cricoid cartilage and the upper border of the first tracheal ring are connected by the _____ membrane.

3. _____

Intrinsic Laryngeal Membranes

1. The intrinsic laryngeal membranes are derived from a broad sheet of connective tissue called the _____ membrane.

1. _____

2. The membrane lining the cone-shaped cavity below the vocal folds is the _____.

2. _____

3. Paired membranes lining the portion of the larynx above the vocal folds are called the _____ membranes. Their superior margins are modified to form the _____ folds.

3. _____
_____

Mucous Membrane of the Larynx

1. The mucous membrane of the larynx is continuous above with the lining of the _____ and below with the lining of the _____.

1. _____
_____

2. The mucous membrane adheres closely to the epiglottis, the aryepiglottic folds and the _____, but is loosely attached in other areas of the larynx.

2. _____

Note: The inhibition of mucus secretion may be a side effect of some drugs.

FIGURE 3.4 LIGAMENTS, MEMBRANES AND JOINTS OF THE LARYNX, ANTERIOR VIEW.

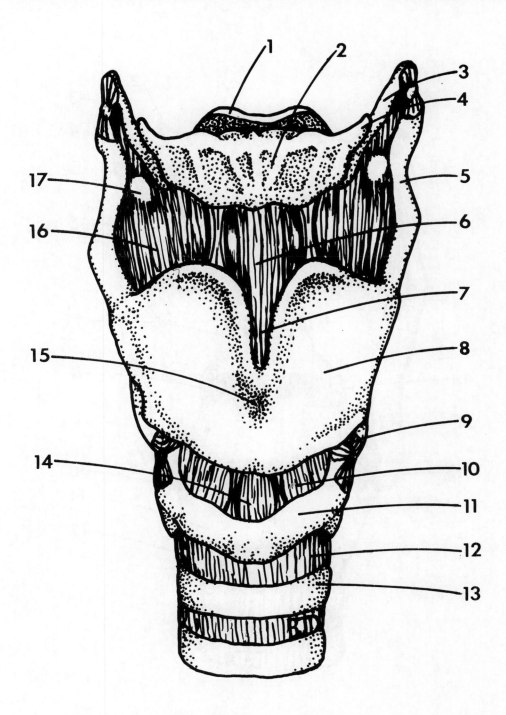

Identify:
1. tip of epiglottis
2. body of hyoid bone
3. greater horn of hyoid
4. lesser horn of hyoid
5. superior horn of thyroid cartilage
6. middle hyothyroid ligament
7. thyroid notch
8. left thyroid lamina
9. cricothyroid articulation
10. conus elasticus
11. cricoid arch
12. cricotracheal membrane
13. first tracheal ring
14. middle cricothyroid ligament
15. thyroid angle
16. hyothyroid membrane
17. hyothyroid foramen

FIGURE 3.5 LIGAMENTS, MEMBRANES AND JOINTS OF THE LARYNX, POSTERIOR VIEW.

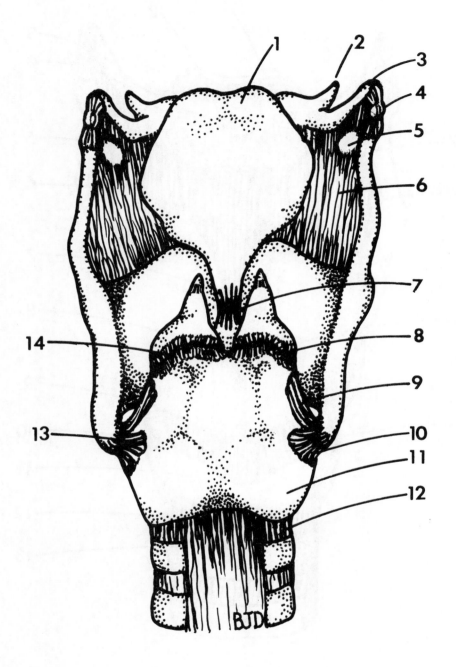

Identify:
1. epiglottis
2. lesser horn of hyoid
3. tip of greater horn
4. triticial cartilage
   (in lateral hyothyroid ligament)
5. hyothyroid foramen
6. hyothyroid membrane
7. thyroepiglottic ligament
8. cricoarytenoid ligament
9. posterior ceratocricoid ligament
10. lateral ceratocricoid ligament
11. posterior cricoid lamina (signet)
12. cricotracheal membrane
13. cricothyroid articulation
14. cricoarytenoid articulation

FIGURE 3.6 LIGAMENTS, MEMBRANES AND JOINTS OF THE LARYNX, IN SAGITTAL SECTION.

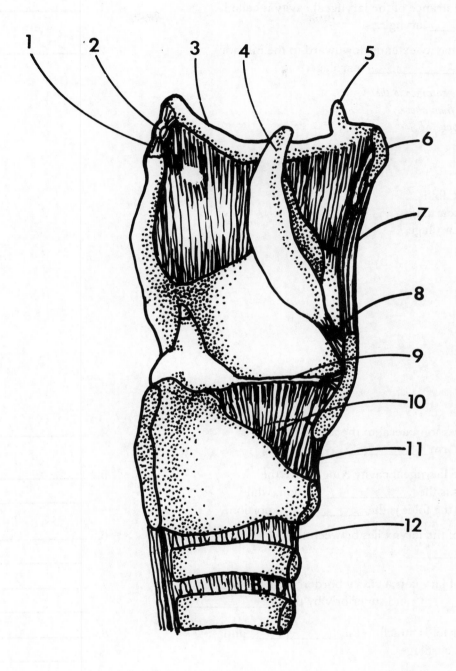

Identify:
1. lateral hyothyroid ligament
2. triticial cartilage
3. greater horn of hyoid bone
4. epiglottis
5. lesser horn of hyoid bone
6. section through hyoid body

7. middle hyothyroid ligament
8. thyroepiglottic ligament
9. vocal ligament
10. conus elasticus
11. middle cricothyroid ligament
12. cricotracheal membrane

No. 3-15 THE INTERIOR OF THE LARYNX: THE CAVITY; THE SUPRAGLOTTAL AND SUBGLOTTAL REGIONS
Text pages 117–118

1. The superior entrance of the laryngeal cavity is called the _____ laryngis.

    1. _____

2. The laryngeal cavity extends downward to the inferior border of the _____ cartilage.

    2. _____

3. *Quickly sketch the entrance to the larynx as viewed from above.* (see Figure 3.3, page 88.)

    Label:
    (1) epiglottis
    (2) aryepiglottic fold
    (3) corniculate cartilage
    (4) cuneiform cartilage
    (5) vocal folds

4. The deep depression lateral to the aditus laryngis is called the pyriform _____.

    4. _____

5. The area of the laryngeal cavity lying above the true vocal folds is the _____ portion, while the area below the folds is the _____ portion.

    5. _____

       _____

6. The vestibule of the larynx lies between the _____ and the _____.

    6. _____

       _____

7. The ventricle of larynx is a cavity bordered inferiorly by the _____ and superiorly by the _____.

    7. _____

       _____

8. The ventricular folds attach to the _____ and _____ cartilages.

    8. _____

       _____

9. The space between the ventricular folds is called the _____ glottis. It is usually *wider / narrower* than the space between the true vocal folds.

    9. _____

       _____

10. Ciliated columnar epithelium lines the _____ region. These cilia beat toward the *trachea / pharynx*.

    10. _____

       _____

    What is their function? _____

No. 3-16 THE INTERIOR OF THE LARYNX: THE VOCAL FOLDS AND THE GLOTTIS
Text pages 118–121

1. *Quickly sketch the vocal folds*
   and glottis as viewed from above
   (See Fig. 3.10, page 102, and
   Figs. 3-29 and 3-30 on text page 119).

   Label:
   (1) anterior commissure
   (2) posterior commissure
   (3) membranous glottis
   (4) cartilaginous glottis
   (5) vocal folds
   (6) thyroid cartilage
   (7) arytenoid cartilage

2. What is the preferred definition of the term glottis?

   2. _____
   _____

3. Each vocal fold is composed of a bundle of muscle tissue
   (_____) and a _____ ligament that is
   continuous with the conus _____.

   3. _____
   _____
   _____

4. The anterior three-fifths of the glottis is known as the
   _____ or _____ glottis.

   4. _____
   _____

5. The posterior two-fifths of the glottis is known as the
   _____ or _____ glottis.

   5. _____
   _____

6. During phonation the most active portion of the vocal folds
   appears to be the *anterior / posterior* portion, which is the
   _____.

   6. _____
   _____

7. The glottis in adult males is usually *more / less* than an
   inch long.

   7. _____

Remember that laryngeal photography produces a mirror image of the vocal folds. Indicate anterior-posterior on the schematics and
photographs in Fig. 3-31A, text page 120.

Quotes: "In textbooks of anatomy, there is no such term as the *posterior commissure*. Some clinicians have used this
term. However, the bilateral vocal folds never meet posteriorly. There is a wall at the posterior end of the
glottis. We believe that the term *posterior commissure* should be eliminated."

"The exact border of the vocal fold laterally and inferiorly has not been clearly defined in any textbooks.
We do not have a good idea either. It is desirable that a reasonable definition be made in the near future."

M. Hirano and S. Kurita in Kirchner, 1986

No. 3-17 THE MUSCLES OF THE LARYNX: INTRODUCTION
Text page 121

Description of laryngeal muscle groups	Extrinsic	Intrinsic
have both attachments within the larynx		✔
have attachments in the larynx and an outside structure	✔	
support and position the larynx	✔	
have predominant control of phonation		✔

1. Most supplemental laryngeal muscles attach to the
    _____ bone. Those elevating the larynx are
    called the _____ muscles, and those
    depressing it are called the _____ muscles.

1. _____
    _____
    _____

No. 3-18 THE EXTRINSIC LARYNGEAL AND FUNCTIONALLY RELATED MUSCULATURE
Text pages 121–126

Muscles: Those positioning and supporting the larynx	Muscle Groups: Extrinsic Tongue	Suprahyoid (↑larynx)	Infrahyoid (↓larynx)	Hyoid Sling	Strap Neck
FUNCTIONALLY RELATED					
digastric		✔		✔	
stylohyoid		✔		✔	
mylohyoid		✔			
geniohyoid		✔		✔	✔
sternohyoid			✔	✔	✔
omohyoid			✔	✔	✔
EXTRINSIC LARYNGEAL					
sternothyroid			✔	✔	✔
thyrohyoid			✔	✔	✔
inferior pharyngeal constrictor					
SUPPLEMENTARY ELEVATORS					
hyoglossus	✔	✔			
genioglossus	✔	✔			

The primary purpose of this chart is to demonstrate that a muscle may be included in more than one functional muscle group.

FIGURE 3.7 EXTRINSIC LARYNGEAL AND FUNCTIONALLY RELATED MUSCULATURE.

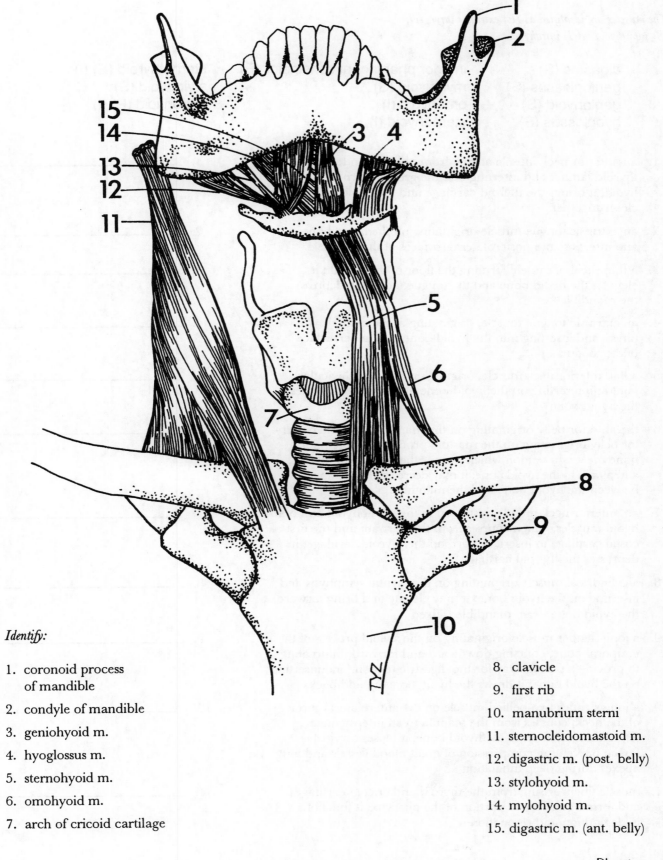

Identify:

1. coronoid process
   of mandible

2. condyle of mandible

3. geniohyoid m.

4. hyoglossus m.

5. sternohyoid m.

6. omohyoid m.

7. arch of cricoid cartilage

8. clavicle

9. first rib

10. manubrium sterni

11. sternocleidomastoid m.

12. digastric m. (post. belly)

13. stylohyoid m.

14. mylohyoid m.

15. digastric m. (ant. belly)

The muscles are identified as (E) extrinsic laryngeal,
(S) suprahyoid or (I) infrahyoid.

digastric (S)	inferior pharyngeal constrictor (E)	sternothyroid (E) (I)
genioglossus (S)	mylohyoid (S)	stylohyoid (S)
geniohyoid (S)	omohyoid (I)	thyrohyoid (E) (I)
hyoglossus (S)	sternohyoid (I)	

1. an anterior neck muscle arising from oblique tendon of the thyroid lamina and inserting into the greater horn of the hyoid; it brings the thyroid cartilage and the hyoid bone closer together

    1. _____

2. an extrinsic tongue muscle originating on the hyoid bone and inserting into posterolateral surfaces of the tongue

    2. _____

3. a thin sheet of muscle forming the floor of the mouth; it elevates the hyoid bone and the tongue, and may depress the mandible

    3. _____

4. an extrinsic tongue muscle, originating at the mental symphysis and inserting into the hyoid bone and undersurface of the tongue

    4. _____

5. a flat anterior neck muscle, originating at sternoclavicular joint and inserting on the hyoid bone; it depresses and fixes the hyoid bone

    5. _____

6. the anterior belly originating on the mandible and the posterior belly originating on the mastoid process of the temporal bone meet at a tendon which pierces the stylohyoid muscle, and attach to the hyoid bone; it may decrease the distance between the hyoid bone and the mandible

    6. _____

7. an anterior neck muscle, just deep to the sternohyoid and omohyoid muscles, which courses from the sternum and the first costal cartilage to insert, in part, on the thyroid cartilage; it depresses the thyroid cartilage

    7. _____

8. a cylindrical muscle originating on the mental symphysis and inserting on the hyoid bone; it may elevate and bring forward the hyoid bone when mandible is fixed

    8. _____

9. a long slender muscle originating on the styloid process of the temporal bone, coursing downward and forward, bifurcating to provide a passageway for the digastric tendon, and inserting on the hyoid bone; it draws the hyoid bone up and back

    9. _____

10. a long slender two-bellied muscle on the anterolateral surface of the neck; courses from the scapula to an intermediate tendon and then on to the hyoid bone; it tenses cervical fascia, preventing compression of great blood vessels and lung apexes during deep inhalation

    10. _____

11. muscle fibers arising from the thyroid and cricoid cartilages and forming the lower portion of the pharynx; it functions in deglutition and resonation

    11. _____

FIGURE 3.8 EXTRINSIC LARYNGEAL MUSCLES AND SOME RELATED STRUCTURES.

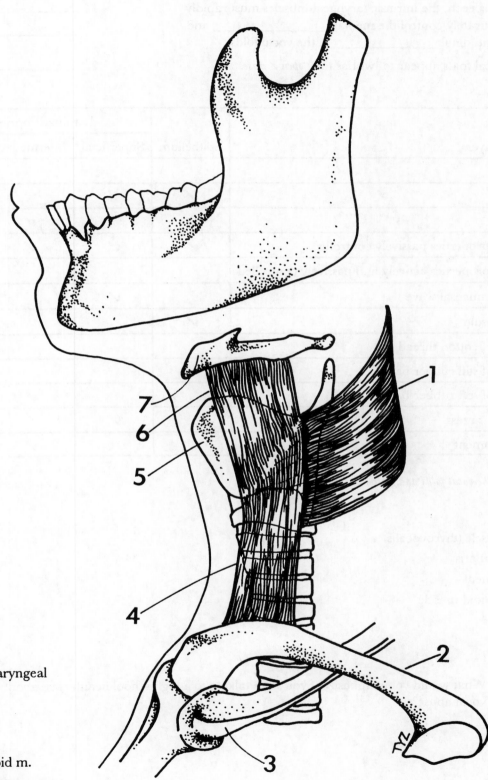

Identify:

1. inferior pharyngeal constrictor
2. clavicle
3. first rib
4. sternothyroid m.
5. thyroid cartilage
6. thyrohyoid m.
7. hyoid bone

1. During speech, the intrinsic laryngeal muscles must rapidly
    and accurately control the medial _____ and
    the longitudinal _____ of the vocal folds.

    1. _____

       _____

2. The vocal folds appear to twist as they *open / close*.

    2. _____

Vocal fold layers	Epithelium	Lamina Propria			Vocalis
		Superficial	Intermediate	Deep	
body					✔
cover	✔	✔			
transition			✔	✔	
mechanical properties passively controlled	✔	✔	✔	✔	
mechanical properties actively and passively controlled					✔
essential for mucosal wave		✔			
an outer capsule	✔				
like bundle of cotton thread				✔	
like bundle of stiff rubber bands					✔
like bundle of soft rubber bands			✔		
like gelatinous mass		✔			
the vocal ligament			✔	✔	

Quickly sketch the vocal folds and glottis as viewed from above.

Label:

(1) vocalis muscle (thyrovocalis)

(2) thyromuscularis

(3) vocal ligament

(4) thyroarytenoid muscle

Question: What would you emphasize if you were talking to a high school health class about
vocal abuse?

FIGURE 3.9 ABDUCTOR AND ADDUCTOR MUSCULATURE AND SOME ASSOCIATED LARYNGEAL STRUCTURES.   (Right thyroid lamina has been removed)

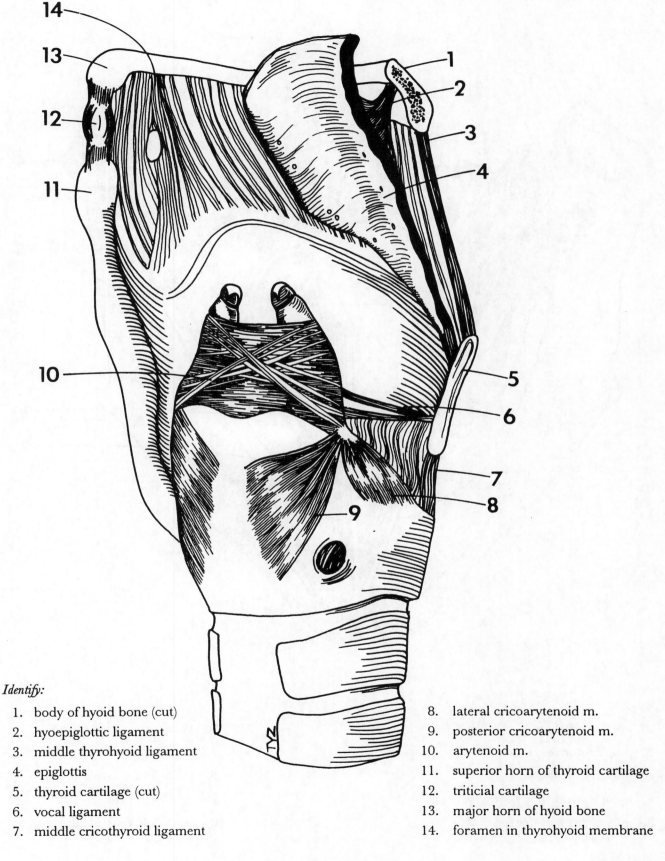

Identify:

1. body of hyoid bone (cut)
2. hyoepiglottic ligament
3. middle thyrohyoid ligament
4. epiglottis
5. thyroid cartilage (cut)
6. vocal ligament
7. middle cricothyroid ligament

8. lateral cricoarytenoid m.
9. posterior cricoarytenoid m.
10. arytenoid m.
11. superior horn of thyroid cartilage
12. triticial cartilage
13. major horn of hyoid bone
14. foramen in thyrohyoid membrane

FIGURE 3.10 **THE LARYNX SHOWN SCHEMATICALLY FROM ABOVE.** Illustrating the partially adducted and twisted left vocal fold and the partially abducted and untwisted right vocal fold. *(The right fold is to your left.)*

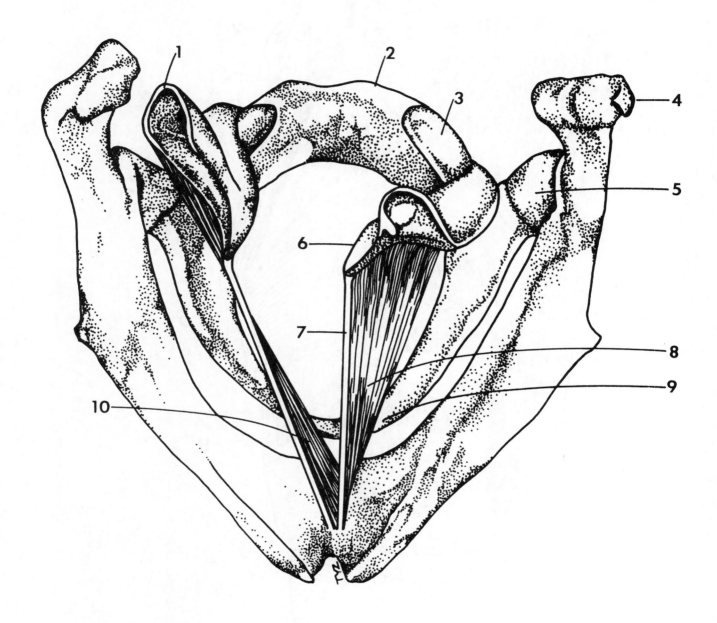

Identify:

1. apex of arytenoid
2. posterior cricoid lamina
3. cricoid articular facet
4. tip of superior horn of thyroid
5. inferior horn of thyroid

6. vocal process of arytenoid
7. vocal ligament
8. vocal fold musculature (left)
9. arch of cricoid
10. vocal fold musculature (right)

No. 3-21 THE MUSCLES OF THE LARYNX: THE INTRINSIC MUSCLES
Text pages 128–136

cricothyroid	oblique arytenoid	thyroarytenoid
lateral cricoarytenoid	posterior cricoarytenoid	transverse arytenoid

1. a broad muscle on the posterior surface of the larynx; a lateral vertical portion appears to function as an abductor, and a medial fan-shaped portion probably fixes the arytenoid cartilages

    1. _____

2. paired muscles which criss-cross the posterior of the paired arytenoid cartilages; contribute to regulation of medial compression

    2. _____

3. a somewhat fan-shaped muscle with a broader origin on the cricoid arch; fibers course up and back, converging to insert on the arytenoid cartilage; an adductor and relaxer

    3. _____

4. the primary muscular mass of the vocal folds; portion flanking vocal ligament is called the vocalis muscle; major regulator of longitudinal tension

    4. _____

5. a fan-shaped muscle arising from the cricoid arch and diverging to insert on the thyroid cartilage as the pars oblique and pars recta; a glottal tensor

    5. _____

6. a muscle coursing horizontally from the postero-lateral surface of one arytenoid cartilage to the other; an adductor

    6. _____

No. 3-22 METHODS OF INVESTIGATION OF LARYNGEAL PHYSIOLOGY
Text pages 137–144

1. List three problems encountered by the developers of laryngoscopy. _____

    _____

    1. _____

    _____

2. A laryngoscope is used for *direct / indirect* examination of the larynx. The use of a mirror permits _____ examination.

    2. _____

    _____

3. A laryngoscope is inserted through the *mouth / nose* and guided into the *esophagus / trachea.*

    3. _____

    _____

4. An instrument which permits observation of moving bodies in such a way that they appear to stop or slow down is called a/an _____.

    4. _____

5. Much of the knowledge of laryngeal function has been derived from the study of _____ motion pictures of the larynx.

    5. _____

6. Lighting up the interior of the larynx by directing a beam of light on the anterior neck is called _____ .

6. _____

7. Why are conventional x-rays of the larynx sometimes unsatisfactory? Why are they particularly unsatisfactory in children?

7. _____

_____

8. The technique for producing radiographs of selected layers of the body is called _____ or _____ . It requires movement of the *x-ray film / x-ray source / both.*

8. _____

_____

_____

9. Strobolaminagraphy, which combines the principles of _____ and _____ , produces _____ films of vocal fold vibration.

9. _____

_____

_____

10. What type of electrodes are used in electromyography of of laryngeal muscles?

Why? _____

10. _____

11. Why is knowledge of anatomy very important for interpretation of EMG findings?

_____

11. _____

12. A pneumotachograph (*pneumo-*, lungs, breath, air; *tacho-* relating to speed) is a device which indicates the quantity or velocity of _____ .

12. _____

13. Measurements of subglottal pressure are usually confined to the research laboratory. Name two techniques used to measure subglottal pressure. Which is most accurate? Why?

13. _____

_____

_____

_____

Note: For direct laryngoscopy adults are usually given a local anesthetic, and children a general anesthetic.

The law of vibrating strings

$$\text{frequency} = \frac{1}{2\ \text{length}}\sqrt{\frac{\text{tension}}{\text{mass}}}$$

length (space between bridges)
tension (weight on string)
mass (thickness of string)

Example
Frequency = 200 Hz

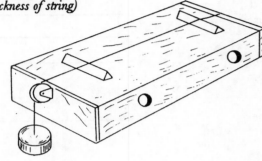

1.

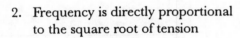

2. Frequency is directly proportional to the square root of tension

function $\propto \sqrt{\text{tension}}$

(tension is increased four times)

Frequency = _____ Hz

3.

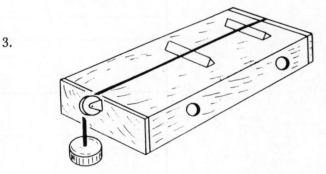

Frequency is inversely proportional to length.

frequency $\propto \dfrac{1}{\text{length}}$

(length is doubled)

Frequency = _____ Hz

Frequency is inversely proportional to the square root of the mass

frequency $\propto \sqrt{\dfrac{1}{\text{mass}}}$

(mass is increased four times)

Frequency = _____ Hz

Note: $\propto$ is the symbol for *is directly proportional to*

Sonninen's equation of vocal fold vibration

$$\text{frequency} = \text{a constant} \times \frac{\text{tension (inner passive + inner active)}}{\text{mass of the vocal fold}}$$

Variables	Pitch	Voice Quality
Frequency of vocal fold vibration affects perceived	✔	✔
Mode (pattern) of vocal fold vibration affects perceived	✔	✔
Configuration of the vocal tract affects perceived	✔	✔

1. The onset of phonation may be divided into the _____ phase and the _____ phase.

    1. _____

       _____

2. The vocal folds move rather abruptly from an abducted position to a fully or partially adducted position during the _____ phase. During this phase subglottal pressure *increases / decreases*, and the velocity of the airstream, as it flows through the restricted glottis, *increases / decreases*.

    2. _____

       _____

       _____

3. Medial compression of the vocal folds is produced by combined action of the adductors, the _____ and _____ muscles, and the abductors, the _____ muscles.

    3. _____

       _____

       _____

4. The extent of medial compression of the vocal folds and the amount of pressure necessary to initiate phonation are *directly / inversely* proportional.

    4. _____

5. The phase that begins with complete or almost complete adduction of the vocal folds and extends through the beginning vibratory cycles is the _____ phase.

    5. _____

6. Greater air flow and air pressure will be required to initiate vocal-fold vibration when the vocal folds are *approximated / almost approximated*.

    6. _____

7. Initial medial movement of partially adducted vocal folds can be accounted for by the _____ effect.

    7. _____

8. If volume fluid flow is constant, its velocity will *increase / decrease* at an area of constriction and its pressure will *increase / decrease* at the constriction.

    8. _____

       _____

9. As air passes through the glottal chink, *positive / negative* pressure will be created between the medial edges of the vocal folds and they will be drawn *together / apart*.

    9. _____

       _____

10. The vocal folds begin to vibrate *before / after* they have actually approximated.

    10. _____

11. Is there general agreement on the importance of the Bernoulli effect in phonation?

    11. _____

12. After the vocal folds have been blown apart, they rebound to an adducted position. List three factors which contribute to adduction of the folds.

    _____

    _____

    _____

The Bernoulli effect may be demonstrated by holding the shorter sides of two half sheets of lightweight paper, one on either side of the mouth, and attempting to blow the free ends apart.

Question: Why do you pucker your lips when you blow out a candle?

No. 3-25 THE ONSET OF PHONATION: THE ATTACK PHASE
Text pages 146–150

Description Types of Attack:	Simultaneous	Breathy	Glottal
expiratory activity significantly precedes laryngeal activity		✔	
respiratory and laryngeal activity begin at approximately the same time	✔		
laryngeal activity precedes respiratory activity			✔
very sudden onset of phonation			✔
considerable release of air before folds are adducted		✔	
generally considered preferable	✔		
weak voice may be a symptom of habitual use		✔	
rough and unpleasant voice quality may be a symptom of habitual use			✔
fricative noise may be superimposed on vocal tone		✔	
generally considered most abusive			✔
vocal folds offer excessive resistance to air flow			✔
vocal folds offer inadequate resistance to air flow		✔	
normal laryngeal function	✔		
laryngeal hyperfunction			✔
laryngeal hypofunction		✔	

Question: Do you agree or disagree with the following statements? Defend your position.

The use of glottal attacks is not necessarily abusive.

In some cases laryngeal hyperfunction is just one manifestation of generally hyper-functional behavior.

High-speed laryngeal cinematography is extremely useful in the diagnosis of voice disorders.

No. 3-26 CHARACTERISTICS OF A VIBRATORY CYCLE
Text pages 150–152

1. What are the three phases of a vibratory cycle? In a typical cycle of vocal fold vibration, which is the longest? Which is the shortest?

    1. _____
       _____
       _____

2. Analysis of high-speed laryngeal cinematography films allows the researcher to study vibratory characteristics of the larynx by graphing glottal _____ as a function of _____.

    2. _____
       _____

3. Stroboscopic illumination synchronized with the rate of vibration of the vocal folds is called _____.

    3. _____

4. _____ quotient = $\dfrac{\text{time of abduction}}{\text{time of adduction}}$

    4. _____

5. _____ quotient = $\dfrac{\text{time the glottis is open}}{\text{total duration of the cycle}}$

    5. _____

No. 3-26 cont'd

6. Which of the preceding quotients generally provides more useful information? Why?

6. _____

_____

7. The vocal fold area that opens first and closes last is the *anterior / posterior* area.

7. _____

8. At conversational pitch and intensity levels, vibration of the vocal folds is mainly along the *vertical / horizontal* plane, and as loudness increases the extent of *vertical / horizontal* displacement increases.

8. _____

_____

9. The lower edges of the vocal folds are the *first / last* to be blown apart and the *first / last* to close. This pattern of vibration produces a _____ phase difference.

9. _____

_____

_____

No. 3-27 THE PITCH-CHANGING MECHANISM: INTRODUCTION
Text pages 152–153

Note: The *mode of vocal fold vibration* refers to the manner or pattern of vocal fold movement. The *mode of pitches produced* refers to the pitch used most frequently.

1. An individual's pitch range during conversational speech is approximately *one / two / three* octave(s).

1. _____

2. The pitch mode, expressed in Hz is called _____.
For young adult males it is approximately _____ Hz.
For young adult females it is approximately _____ Hz.

2. _____

Male _____ Female _____

3. Pitch mode, expressed in musical notes, is usually called pitch _____.

3. _____

4. Normally the distributions of pitch ranges are *positively / negatively* skewed.

4. _____

5. Why may some infrequent, very low pitches that occur during speech not be included in the pitch range?

5. _____

6. The pitch level at which a person *should* speak, which is primarily determined by anatomical and physiological characteristics, is called the _____ or _____ pitch level.

6. _____

_____

7. The pitch level at which a person *does* speak is called the _____ pitch level.

7. _____

8. In superior male speakers, optimum pitch level is usually ¼ / ⅓ / ½ the way up the singing range.

8. _____

Note: Someone speaking at an inappropriate pitch level may use a more natural pitch level when coughing, throat-clearing, grunting while lifting, or laughing spontaneously.

1. Vocal folds in the abducted position are considerably *shorter / longer* than they are during phonation.

   1. _____

2. As pitch increases

   2.

   a. the length of the vocal folds _____.

   a. _____

   b. the mass or cross-sectional area of the folds _____.

   b. _____

3. According to the law of vibrating strings, as pitch increases

   3.

   a. length of the string should _____.

   a. _____

   b. mass of the string should _____.

   b. _____

4. Are changes in vocal fold mass sufficient to fully account for changes in pitch?

   4. _____

5. It is possible that pitch increases may be due entirely to changes in _____ of the vocal folds.

   5. _____

6. The three intrinsic laryngeal muscles primarily responsible for pitch increases are the two glottal tensors, the _____ and the _____, and the glottal abductor, the _____.

   6. _____
   _____
   _____
   _____

7. As pitch increases the two glottal tensors function as *agonists / antagonists*, and the glottal abductor anchors the _____ cartilage.

   7. _____
   _____

Function	Thyroarytenoid	Cricothyroid
without antagonistic muscular forces, contraction primarily shortens the folds	✔	
without antagonistic muscular forces, contraction lengthens the folds		✔
probably responsible for external tension or "loading" of the vocal folds		✔
probably responsible for internal tension or minute changes and adjustments within the vocal folds	✔	

Description Pitch level:	Natural	High
folds appear stiff and rigid		✔
folds are rounded and thick	✔	
folds seem relaxed, almost flaccid	✔	
glottis appears rhomboid-shaped	✔	
glottis appears as a variable slit		✔
only medial edges of the folds seem to vibrate		✔
folds appear to vibrate as a whole	✔	
folds are elevated and tilted		✔

No. 3-28 cont'd

8. At extremely high pitches the voice tends to become *harsh / breathy*.

   Why? _____

   8. _____

9. As pitch increases, the thickness and cross-sectional area of the folds decreases, with greater changes occurring in the *high, low* portion of the pitch range.

   9. _____

10. Glottal _____ or _____ = $\dfrac{\text{Subglottal pressure}}{\text{Rate of airflow}}$

    10. _____

    _____

11. If laryngeal tension is held constant, an increase in subglottal pressure will usually produce a/an *significant / insignificant* increase in pitch.

    11. _____

12. For normal speakers, glottal resistance is lowest at *low / habitual / high* pitch levels.

    12. _____

With your hands representing the vocal folds, demonstrate vocal fold tilt.

No. 3-29 THE PITCH-LOWERING MECHANISM; EXTRINSIC LARYNGEAL MUSCLES AND PITCH CHANGE
Text pages 159–160

1. Habitual pitch is near the *upper / lower* limits of the pitch range.

   1. _____

2. The lowering of vocal pitch is due to an increase in _____ and/or a decrease in _____.

   2. _____

   _____

3. Glottic margins can be relaxed by tissue _____ and the folds can be shortened by contraction of the _____ muscle.

   3. _____

   _____

4. At low pitch levels, medial compression is facilitated by the _____ muscle.

   4. _____

   _____

5. In what regions of a person's pitch range are the extrinsic laryngeal muscles more likely to be active?

   5. _____

6. Any changes in the larynx are the result of the _____ sum of the various forces in action.

   6. _____

Quote: "Optimal pitch range is that area within the total extended pitch of the voice where maximum amount of voice is produced with a minimum amount of effort."

Morton Cooper, 1971 (Travis)

Muscles	Function:				Influence:	
	Abduct	Adduct	Tense	Relax	Medial Compress.	Long. Tension
oblique arytenoid		✔			✔	
transverse arytenoid		✔			✔	
post. cricoarytenoid	✔				✔	
lat. cricoarytenoid		✔		✔	✔	
thyroarytenoid		✔	✔	✔		✔
cricothyroid			✔			✔

Length, tension, and mass (per unit length) of the vocal folds may be modified simultaneously. Indicate the effects of contraction of individual muscles by labeling the arrows in the following schematic.

Abbreviations:

CTH cricothyroid
LCA lateral cricoarytenoid
OBA oblique arytenoid
PCA posterior cricoarytenoid
THA thyroarytenoid
TRA transverse arytenoid

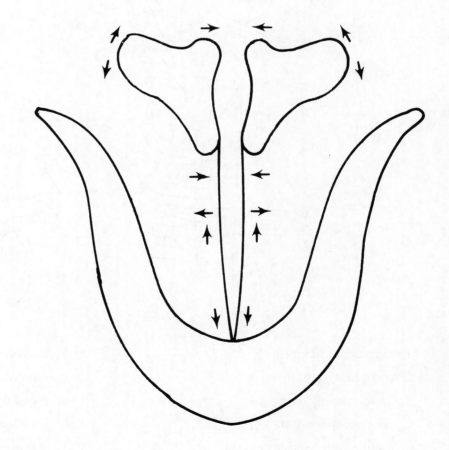

In the following schematic of the actions of the extrinsic laryngeal and functionally related musculature, the arrows suggest the consequences of individual muscle contraction.

Attempt to generate the names of muscles. Speculate on the consequences of contraction of single muscles and of groups of muscles.

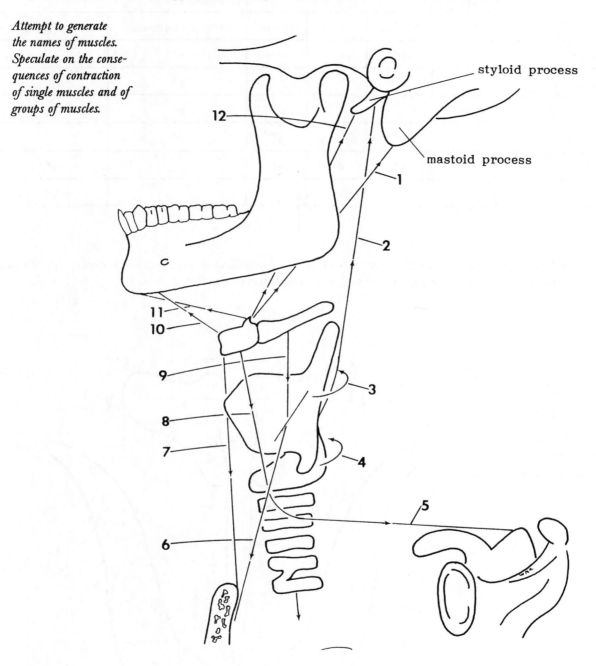

styloid process

mastoid process

Key	1. digastric (posterior belly) m.	7. sternohyoid m.
	2. stylopharyngeus m.	8. omohyoid (anterior belly) m.
	3. thyropharyngeus m.	9. thyrohyoid m.
	4. cricopharyngeus m.	10. geniohyoid m.
	5. omohyoid (posterior belly) m.	11. digastric (anterior belly) m.
	6. sternothyroid m.	12. stylohyoid m.

No. 3-32 THE INTENSITY-CHANGING MECHANISM
Text pages 160–164

1. An individual's total range of vocal intensity is approximately *70 / 90 / 110* decibels, while the range during conversational speech is approximately *20 / 30 / 40* decibels.

1. _____
_____

2. Intensity at high pitch levels in controlled primarily by *rate of air flow / glottal resistance*. Why?

2. _____
_____

3. At low pitch levels less air expenditure may be required for high intensity phonation than for low intensity phonation. Defend your answer.

3. T F
_____

4. Rate of Air Flow = $\dfrac{\text{Subglottal Pressure}}{\text{Glottal Resistance}}$

4.

   a. The rate of air flow is directly related to _____.

   a. _____

   b. The rate of air flow is inversely related to _____.

   b. _____

You are speaking and suddenly increase your vocal intensity. Predict probable effects on	Increase	Decrease	No Change
the duration of the closed phase of the vibratory cycle	✔		
the maximum glottal area			✔
the forces of medial compression	✔		
glottal resistance (impedance)	✔		
subglottal pressure	✔		
pitch	✔ very slight		
activity of the adductor muscles	✔		
air flow	✔		

Note: Disorders that prevent closure of the vocal folds or that obstruct or restrict the respiratory passages may reduce vocal intensity, but there is usually no organic basis (except deafness) for excessive vocal intensity. Speaking too loudly, particularly if habitual pitch is not at the natural level, may contribute to the development of laryngeal pathology.

No. 3-33 TRANSGLOTTAL PRESSURE DIFFERENTIAL
Text pages 164–165

1. Transglottal Pressure Differential equals _____ pressure minus _____ pressure.

1. _____
_____

2. Total Resistance to Air Flow in the Speech Mechanism equals _____ resistance plus _____ resistance.

2. _____
_____

No. 3-33 cont'd

3. As you begin to produce the sound [b] in the word "Abe":   3.

    a. supraglottal pressure will *increase / decrease.*   a. _____

    b. transglottal pressure differential will *increase / decrease.*   b. _____

    c. the pharynx will *expand / contract.*   c. _____

4. As vocal tract resistance increases, checking action *increases /*   4. _____
decreases, thus demonstrating a/an *direct / inverse* relationship
between vocal tract resistance and checking action.   _____

No. 3-34 VOICE REGISTERS, THE LIMITS OF PITCH RANGE, THE VIBRATO (VOCABULARY)
Text pages 165–171

damping	loft	tremolo
falsetto	modal pitch range	trill
fundamental frequency	pitch	vibrato
glottal fry	pulse register	
laryngeal whistle	register	

1. small and rapid pitch and intensity changes during singing   1. _____

2. a portion of the pitch range which is similar in production
and quality, and which differs from other portions   2. _____

3. vocal fold vibration expressed in terms of musical notes,
e.g., A_4 3. _____

4. extreme upper portion of pitch range (2)   4. _____

    _____

5. exaggerated vibrato   5. _____

6. normal pitch range   6. _____

7. causing a decrease in amplitude of successive waves or
oscillations   7. _____

8. vocal fold vibration expressed in cycles per second, e.g.,
440 Hz   8. _____

9. flutelike sound produced by escape of air between non-
vibrating vocal folds   9. _____

10. in music, a rapid alternation of two notes a whole or half
tone apart   10. _____

11. a popping or creaky sound produced by phonating quietly at
the lowest possible pitch (2)   11. _____

    _____

No. 3-35 THE LIMITS OF THE PITCH RANGE
Text pages 167–169

1. As a singer goes from one register to another, does the
   mode of vocal fold vibration change?

    1. _____

2. Why do lower pitched voices generally sound "richer" than
   higher voices?

    2. _____

_____

Description	Falsetto	Pulse Register (Glottal Fry)
air bubbles out of the larynx in discrete bursts		✔
vocal folds come together only at the very edges	✔	
vocal folds are fully approximated		✔
low rate of air flow		✔
long closed phase during vibratory cycle		✔
single and double patterns of vibration (syncopated rhythm)		✔
glottal chink may be quite large	✔	

Note: Pulse register, also called glottal fry, vocal fry, or creaky voice, is a part of the normal vocal repertory, but excessive use may indicate that a person's habitual pitch level is too low.

No. 3-36 VOICE QUALITY; SPECIFIABLE PARAMETERS OF VOICE PRODUCTION
Text pages 171–176

1. Maximum frequency (pitch) range usually encompasses
   1 / 2 / 3 octaves.

    1. _____

2. The mean rate of vocal fold vibration specifies _____ pitch.

    2. _____

3. Maximum phonation time specifies _____ cost.
   An adult speaker can comfortably sustain phonation for
   approximately *15–25 / 25–35 / 35–45* seconds.

    3. _____
   _____

4. Minimum-maximum intensity at various pitches is measured
   with a/an _____ meter. The minimum-maximum SPL
   in frequency midrange is approximately *40 / 50 / 60* dB.

    4. _____
   _____

5. Slight cycle-to-cycle differences in the vibratory period
   are called _____.

    5. _____

   a. As cycle-to-cycle variations increase, the voice
      will sound increasingly _____.

    a. _____

   b. The same frequency variations in vibratory period
      will generally have a more negative impact on the
      quality of a *male / female* voice.

    b. _____

6. Although judges may not agree that a particular voice is
   harsh or hoarse or breathy or rough, the differences they
   perceive may all be called _____.

    6. _____

7. Although nasality is often called a voice disorder, it is
   actually a/an _____ problem.

    7. _____

Description Vocal Quality:	Breathy	Harsh
glottal attack initiates phonation		✔
poor approximation of folds generates frictional noise	✔	
aperiodic vocal fold vibration		✔
lateral excursion of folds may be excessive	✔	
tension and medial compression of folds may be excessive		✔
posterior glottal chink may be present	✔	
limited intensity	✔	

Description nasality:	Hyper-	Hypo-
inadequate coupling of nasal passages to oral and pharyngeal cavities		✔
excessive coupling of nasal passages to oral and pharyngeal cavities	✔	
associated with cleft palate	✔	
associated with very enlarged adenoids		✔
due to inadequate velopharyngeal closure	✔	
sounds "stuffed-up"		✔

Quote: "Two interesting words that have made their way to the speech and voice science literature are *jitter* and *shimmer*. They refer to short-term (cycle-to-cycle) variability in fundamental frequency and amplitude, respectively. . . . An unfortunate misunderstanding can arise for singing teachers who use the term *shimmer* to describe a beautiful bell-like vocal quality. A shimmering voice is aesthetically most pleasing in this context. As a short-term amplitude perturbation, however, shimmer is not particularly pleasing. It is usually perceived as a crackling or buzzing, and in extreme cases, it can become very unpleasant and rough."

Ingo R. Titze, 1994

No. 3-37 WHISPER
Text pages 176–177

Description	Whisper	Norman Phonation
sound generated by vibration		✔
sound generated by friction	✔	
sounds are aperiodic	✔	
sounds have a fundamental frequency and a harmonic structure		✔
rate of air flow is very high	✔	
arytenoid cartilages tend to "toe-in"	✔	
medial surfaces of the arytenoid cartilages meet when the folds come together		✔
"nonvocal sound" production	✔	

1. Does whispering appear to be abusive to the vocal folds?   1. _____

   Why? _____

Description Larynx:	Infant	Adult
lower border about level with C7		✔
lower border about level with C2-3	✔	
soft, pliable hyaline cartilage	✔	
rounded junction of thyroid laminae	✔	
angular junction of thyroid laminae		✔
superficial layer (mucosa) of lamina propria predominates	✔	
intermediate and deep layers (vocal ligament) of lamina propria are well developed		✔
very close to hyoid bone	✔	
proportionally small arytenoid cartilages		✔
short and wide	✔	
no observable sex differences	✔	

Description Larynx:	Male	Female
undergoes mutation during puberty	✔	
grows during puberty	✔	✔
cartilaginous structure grows rapidly during puberty	✔	
rather rounded junction of thyroid laminae		✔
slightly smaller thyroid angle	✔	
descends throughout life	✔	✔
thyroid and cricoid cartilages begin to ossify in early adulthood	✔	✔
elastic cartilages become partially calcified in later years	✔	✔
hyaline cartilage usually ossified by age 65	✔	✔

1. The length of the vocal folds almost doubles during the first *year / five years / fifteen years*.

    1. _____

2. Why, during puberty, does the lower range of the voice drop about an octave in males and only a few notes in females? _____

    2. _____

3. Is there a significant relationship between the angle of the thyroid and length of the vocal folds? Explain.

    3. _____

4. As the larynx grows, pitch level tends to *lower / rise* and at middle age it tends to *lower / rise* slightly.

    4. _____

5. During the aging process, pitch range may tend to *increase / decrease*. Why?

    5. _____

The spectrograms of speech samples on the opposite page include one normal production and six productions which are in one way or another, departures from the normal production. Three spoken words; *heed, hid, had,* were used as samples. Specifiable parameters of the voice include periodicity of vocal fold vibration, the presence or absence of noise or turbulence (excluding consonants), the rate of vocal fold vibration and the presence or absence of voicing. (Whisper is an example of speech production without vocal fold vibration.)

Each vertical striation represents one burst of air through the vibrating vocal folds. Striations spaced closer together indicate a higher rate of vocal fold vibration while striations spaced further apart indicate a lower rate of vibration. Note in the normal sample the fairly regular intervals between the vertical striations, or in other words, *periodic vocal fold bursts.*

	The following samples were produced:	*Circle no. of sample described*
A.	with a very low vibratory rate	1 2 3 4 5 6
B.	with an irregular or aperiodic vocal fold rate (jitter)	1 2 3 4 5 6
C.	without vocal fold vibration, or in other words whisper. The spectrogram indicates noise or turbulence throughout the entire production and throughout the frequency range	1 2 3 4 5 6
D.	with fairly periodic vocal fold vibration, but with considerable turbulence or noise superimposed on the voice	1 2 3 4 5 6
E.	with excessive nasality, thus resulting in a loss of definition of the horizontal dark bands which represent the vocal tract resonances or formants	1 2 3 4 5 6
F.	with a very high rate of vocal fold vibration	1 2 3 4 5 6
G.	with a breathy voice	1 2 3 4 5 6
H.	with pulse register (glottal fry)	1 2 3 4 5 6
I.	with a high-pitched voice	1 2 3 4 5 6
J.	with a harsh voice	1 2 3 4 5 6
K.	with hypernasality	1 2 3 4 5 6
L.	with whispered speech	1 2 3 4 5 6

Note: Shimmer (cycle-to-cycle variations in amplitude) is another specifiable parameter of voice production.

Quote: "... in both spectral and fundamental frequency analysis, new mathematical approaches and new ways of displaying the signal and its components have been appearing regularly. The excitement of speech science today is not only in new understanding of speech and new practical applications of that understanding, but also in new ways to gain understanding. Similarly, the motivation for all this development has at least three sources: the basic desire to understand a central human activity, the desire to develop better therapy for speech that has somehow gone wrong, and considerable commercial interest in speech synthesis and speech recognition. In particular, the difficulty of programming machines to recognize speech has forced us to recognize that what we knew about speech even five years ago was incomplete."

R. Kent and C. Read, 1992

8000

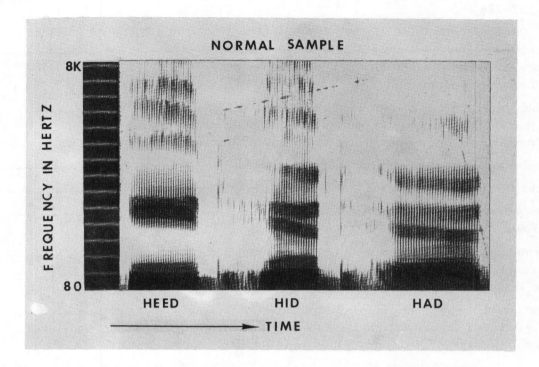

NORMAL SAMPLE

FREQUENCY IN HERTZ

8K

80

HEED   HID   HAD

TIME

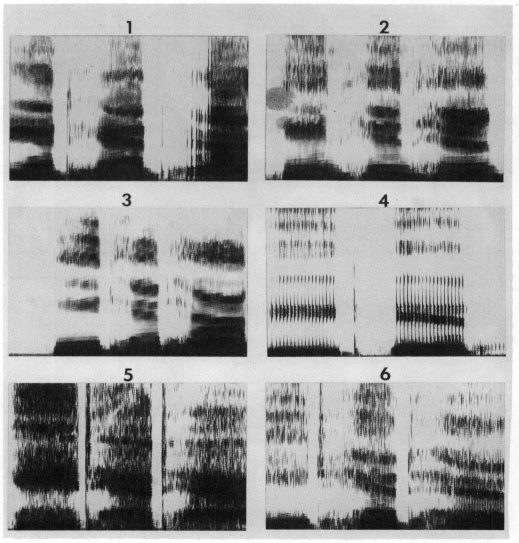

1. Which theory contends that vocal fold vibration is almost totally dependent on the rate of neural impulses received by the laryngeal muscles?

1. _____

2. Which theory contends that vocal fold vibration depends on the laws of physics and that the physical properties of the larynx are regulated, for the most part, by the intrinsic laryngeal muscles?

2. _____

3. Are the two theories generally considered incompatible?

3. _____

4. Which theory is widely accepted?

4. _____

5. The earliest models of the larynx and vocal tract were *mechanical / mathematical.*

5. _____

6. Attempting to duplicate vowel sounds by first analyzing the complex wave form and then simultaneously producing all of its components is called speech _____.

6. _____

7. Digital computer simulation of speech is produced by using *mechanical / mathematical* models of the larynx. What are some of the advantages of this type of model?

7. _____

_____

_____

8. A single-degree-of-freedom model of the larynx, representing the vocal folds as a mechanical _____, accounts solely for movement of the vocal fold mass to and from the _____.

8. _____

_____

9. In addition to the vocal fold movement accounted for in a single-degree-of-freedom model, the two-degree-of-freedom model also accounts for _____ phase differences.

9. _____

10. How do the mucoviscoelastic-aerodynamic and the myo-elastic-aerodynamic theories of vocal fold vibration differ?

10. _____

_____

_____

11. In Titze's sixteen-mass model each vocal fold is divided into two portions, the _____ membrane and the _____ muscle-_____ ligament mass.

11. _____

_____

_____

12. Titze's entire model, which includes, not only the vocal folds, but also, approximations of the pharynx, mouth and the nasal tract, permits study of both phonation and _____.

12. _____

Quote: "Two avenues of investigation likely to be pursued with vigor over the next several years are application of fluid mechanics and its recent offspring, deterministic chaos, to the study of phonatory processes. Work in these areas may well lead to enhanced mathematical models, to an improved ability to synthesize natural-sounding speech, and ultimately to a greater understanding of normal and abnormal phonatory function."

R. Orlikoff and J. Kahane, in Lass, 1996

Chapter 4
Articulation

No. 4-1 ARTICULATION: INTRODUCTION
Text page 198

1. Short duration vibrations generated within the supraglottal air column constitute the glottal or laryngeal _____.

2. Any physical component of a complex tone is called a/an _____.

3. Prominent resonances of the vocal tract are called _____.

4. Speech sounds may be generated without vibration of the vocal folds.

1. _____

2. _____

3. _____

4. T F

No. 4-2 THE BONES OF THE SKULL (VOCABULARY)
Text pages 198–203

calvaria	frons	occiput	temple
cranium	lacri-	orbit	turbinate
facial skeleton	meatus	septum	vertex

1. most superior point of the skull

2. a passageway or canal, particularly the external opening

3. two major divisions of the skull

4. the back of the skull

5. pertaining to tears

6. scroll-like, spiraled

7. houses and protects the brain

8. forehead

9. lateral side of the forehead

10. bowl comprising the skull cap

11. bony depression housing the eye

12. a partition separating two cavities

1. _____

2. _____

3. _____
   _____

4. _____

5. _____

6. _____

7. _____

8. _____

9. _____

10. _____

11. _____

12. _____

concha	dental alveoli	pterygoid
condyle	lingula	ramus
coronoid	mental	semilunar
corpus	mylo-	symphysis

1. a. resembling a half-moon or crescent

   b. tooth socket

   c. shaped like a wing

   d. a branch

   e. body

   f. a rounded projection on a bone

   g. pertaining to chin

   h. small tongue

   i. shaped like a beak or a crown

   j. a site or line of union between two structures

   k. shell-like

   l. pertaining to mill, molar

2. The point at which the two halves of the mandible are joined is called the _____ symphysis.

3. The superior surface of the U-shaped portion of the mandible is called the _____ arch.

4. Is the condylar process anterior or posterior to the coronoid process?

5. How does the mandible contribute to speech production?

1. a. _____

   b. _____

   c. _____

   d. _____

   e. _____

   f. _____

   g. _____

   h. _____

   i. _____

   j. _____

   k. _____

   l. _____

2. _____

3. _____

4. _____

5. _____
   _____
   _____
   _____

Quote: "A skull in the hand is worth two in the book."

Anonymous

Palpate your own mandible, identifying angles and the sites of landmarks. You may also want to try this on a <u>willing</u> subject.

No. 4-4 BONES OF THE FACIAL SKELETON: THE MAXILLAE
Text pages 207–209 *Study figures in text.*

1. The maxillae form the
   a. _____ jaw.
   b. floor and lateral walls of the _____ cavity.
   c. floor of the _____ cavity.

2. The maxillae also contribute to the formation of the _____ of the mouth.

3. The body of the maxilla is shaped somewhat like a _____.

4. Name the four processes of each maxilla.

5. Why do the maxillae need the support of buttresses of bone?

6. The anterior termination of the midline suture of the palatine processes of the maxillae is at the _____ foramen.

7. The triangular portion of bone which contains the sockets for the upper incisors is sometimes called the _____.

8. Which two bones of the facial skeleton are not paired?

1. _____
   a. _____
   b. _____
   c. _____

2. _____

3. _____

4. _____
   _____
   _____
   _____

5. _____
   _____

6. _____

7. _____

8. _____
   _____

Note: In the early stages of embryological development the premaxilla (premaxillary process) and the palatine part of the maxillary process are separate entities. Normally they fuse during the eighth week. Cleft palate will result when the juncture is absent or incomplete.

No. 4-5 BONES OF THE FACIAL SKELETON: NASAL, PALATINE, LACRIMAL, ZYGOMATIC, INFERIOR NASAL CONCHAE, VOMER
Text pages 209–213 *Study figures in text.*

1. The nasal bones form the _____ of the nose.

2. The posterior fourth of the hard palate is formed by the _____ bones.

3. The posterior nasal spine is formed by the _____ bones.

4. The smallest facial bones are the _____ bones.

5. The bone which contributes to the formation of the lateral walls and floor of the orbital cavity is the _____.

6. The inferiormost portion of the lateral nasal wall is made up of the inferior _____, also called the inferior _____ bones.

7. The lower half of the bony nasal septum is formed by the _____. Its anterior portion articulates with the _____ nasal septum.

1. _____

2. _____

3. _____

4. _____

5. _____

6. _____
   _____

7. _____
   _____

No. 4-6 THE BONES OF THE CRANIUM: VOCABULARY
Text pages 213–223

antrum	crista galli	labyrinth	parietal	sphenoid
cecum	ethmoid	lambdoid	petrous	squama
chiasma	glabella	lamellar	pterygoid	styloid
clinoid	glosso-	mastication	raphe	temporal
coronal	hamulus	mastoid	rostrum	tympanic
cribriform	hypophysis	nuchal	sella	zygo-

1. pertaining to crown
2. pillarlike; long and pointed
3. pertaining to yoked
4. a seam or ridge indicating the line of union of two symmetrical halves
5. pituitary gland; a downgrowth
6. pertaining to time
7. wing-shaped
8. cock's comb
9. forming or situated on a wall
10. area between the eyebrows
11. a saddle-shaped depression
12. a cavity or hollow space, especially in bone
13. perforated or sievelike (2)

14. chewing
15. resembling a bed
16. a beaklike appendage or a platform
17. pertaining to tongue
18. derived from a word meaning drum
19. wedge-shaped
20. shaped like the Greek letter λ or Λ
21. arranged in thin plates
22. a thin plate or scale of bone
23. an intricate maze of connecting pathways
24. hard, stony
25. a blind pouch
26. breastlike
27. pertaining to the back of the head
28. a hook-shaped process
29. shaped like the letter X

1. _____
2. _____
3. _____
4. _____
5. _____
6. _____
7. _____
8. _____
9. _____
10. _____
11. _____
12. _____
13. _____
   _____
14. _____
15. _____
16. _____
17. _____
18. _____
19. _____
20. _____
21. _____
22. _____
23. _____
24. _____
25. _____
26. _____
27. _____
28. _____
29. _____

Note: The term *clinical* originally pertained to the bedside of a patient. The words *clinical, laminate, timpani, petrify, mastectomy, pterodactyl,* and *stylus* are etymologically related to some of the words in this exercise.

No. 4-7 THE BONES OF THE CRANIUM
Text pages 213–223 *Study figures in text.*

The Ethmoid Bone

1. The cribriform plate contributes to the roof of the
_____ cavity and separates it from the _____
cavity.

    1. _____

    _____

2. The perforations of the cribriform plate allow for
passage of the _____ nerves into the _____
cavity.

    2. _____

    _____

3. Two scroll-like extensions of the ethmoid labyrinth
form the middle and superior nasal _____.

    3. _____

*Pay particular attention to Figure 4-18, text page 214, which illustrates
the articulations of the perpendicular plate.*

The Frontal Bone

1. The vertical plate or squamous portion of the bone forms
the _____.

    1. _____

2. The horizontal plate contributes to the roof of the
_____ and _____ cavities.

    2. _____

    _____

3. The opening occupied by the cribriform plate is called
the _____ notch.

    3. _____

The Parietal Bones

1. The parietal bones form most of the rounded roof of the
_____.

    1. _____

Sutures Join:	Parietal	Temporal	Frontal	Occipital
Coronal	✔		✔	
Squamosal	✔	✔		
Lambdoid	✔			✔
Sagittal	✔			

The Occipital Bone

1. The occipital bone forms the lower and back portion of the
_____.

    1. _____

2. The demarcation between the brain and the spinal cord
occurs at the _____.

    2. _____

3. The first cervical vertebra articulates with the
occipital _____.

    3. _____

The Temporal Bones

1. The temporal bones form the lateral base and sides of the _____ .

2. The most conspicuous landmark of the squamous portion is the _____ process.

3. The organs of hearing and equilibrium are located in the _____ portion.

4. The mastoid process is characterized by the presence of _____ cells which communicate, directly or indirectly, with the middle ear cavity.

5. The middle ear cavity is in the _____ portion.

6. Muscles inserting in the pharynx, tongue, and on the hyoid bone originate on the _____ process.

7. How is the mandible joined to the temporal bones?

1. _____

2. _____

3. _____

4. _____

5. _____

6. _____

7. _____
_____

Question: We know that temporal refers to time. Can you guess why this bone was named the "time" bone? It has something to do with the aging process.

The Sphenoid Bone

1. What are the main parts of this bone?

1. _____
_____
_____
_____

2. The sphenoid bone is located at the base of the skull between the _____ and _____ bones.

2. _____
_____

3. The sphenoid sinuses are in what part of the bone?

3. _____

4. The bony septum of the nasal cavity is formed, in part, by the perpendicular plate of the _____ bone articulated with the midline crest of the sphenoid bone.

4. _____

5. The sella turcica houses the _____ .

5. _____

6. What part of the sphenoid bone partially houses the eye?

6. _____

7. What part of the sphenoid bone is divided into a medial and lateral plate or lamina?

7. _____

8. Why are the spine of the sphenoid and the pterygoid hamulus particularly important in the study of speech mechanism?

8. _____
_____

9. Why is the sphenoid bone sometimes called the "keystone" of the skull?

9. _____
_____

FIGURE 4.1 THE SKULL AS SEEN FROM THE FRONT.

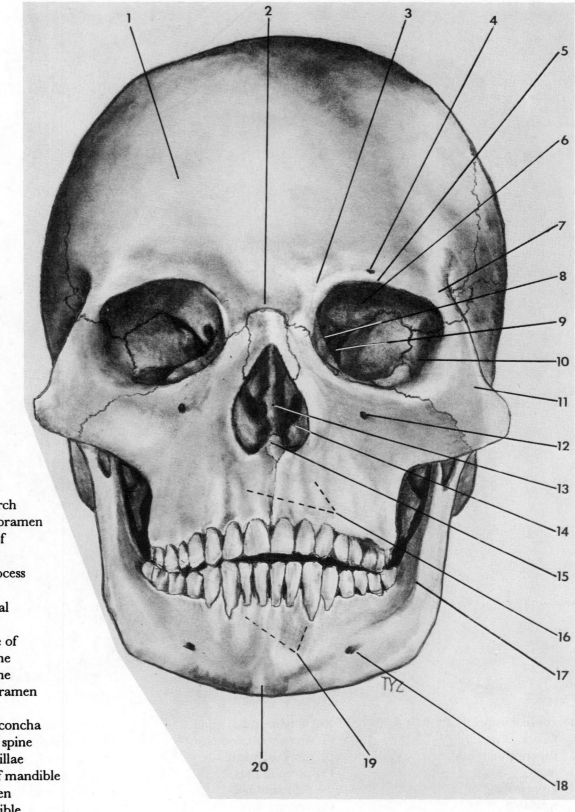

Identify:

1. frontal bone
2. glabella
3. superciliary arch
4. supraorbital foramen
6. orbital plate of frontal bone
7. zygomatic process
8. optic canal
9. superior orbital fissure
10. orbital surface of zygomatic bone
11. zygomatic bone
12. infraorbital foramen
13. nasal septum
14. inferior nasal concha
15. anterior nasal spine
16. bodies of maxillae
17. oblique line of mandible
18. mental foramen
19. body of mandible
20. mental protuberance

FIGURE 4.2 THE SKULL AS SEEN FROM THE SIDE.

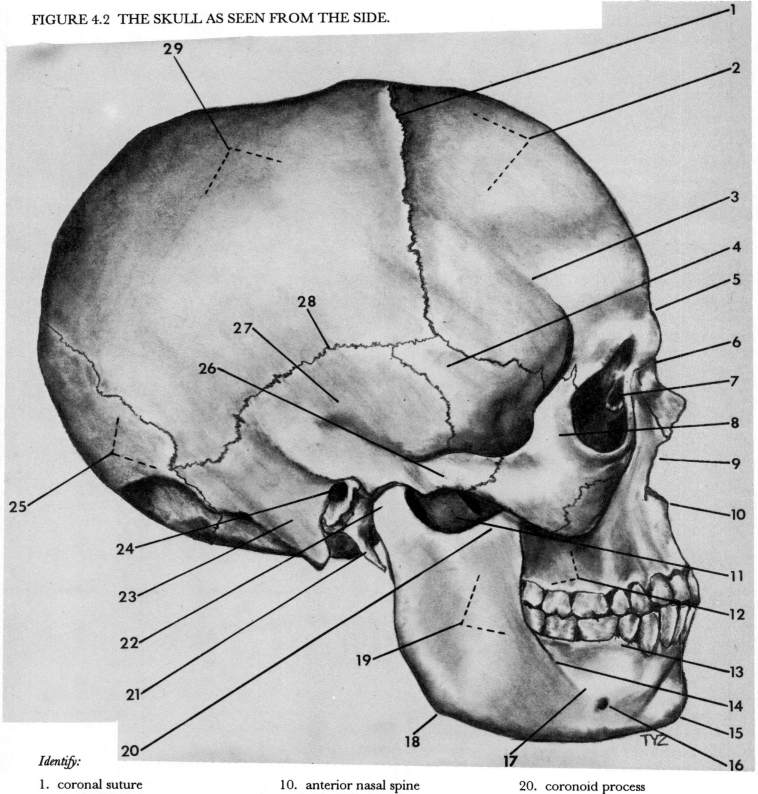

Identify:

1. coronal suture
2. frontal bone
3. superior temporal line
4. greater wing of sphenoid
5. superciliary notch
6. nasal bone
7. lacrimal bone
8. zygomatic bone
9. nasal notch

10. anterior nasal spine
11. mandibular notch
12. body of maxilla
13. alveolar part (of mandible)
14. oblique line
15. mental protuberance
16. mental foramen
17. body of mandible
18. angle of mandible
19. ramus of mandible

20. coronoid process
21. styloid process
22. condylar process
23. mastoid process
24. external auditory meatus
25. occipital bone
26. zygomatic arch
27. temporal bone
28. squamosal suture
29. parietal bone

FIGURE 4.3 THE SKULL AS SEEN FROM BENEATH.

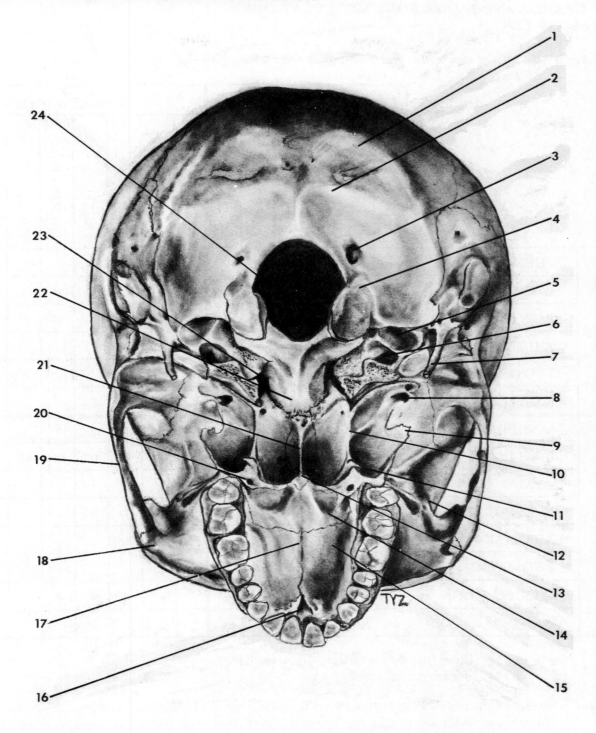

Identify:

1. superior nuchal line
2. inferior nuchal line
3. condylar fossa
4. occipital condyle
5. jugular canal
6. carotid canal
7. styloid process
8. foramen ovale
9. lateral pterygoid plate
10. medial pterygoid plate
11. hamulus of medial pterygoid plate
12. greater wing of sphenoid bone
13. posterior nasal spine
14. horizontal part of palatine bone
15. palatine process of maxilla
16. incisive foramen
17. median palatine (intermaxillary) suture
18. zygomatic bone
19. zygomatic arch
20. lesser palatine foramina
21. vomer bone
22. basilar part of sphenoid bone
23. foramen lacerum
24. foramen magnum

No. 4-8 CRANIAL AND FACIAL BONE ARTICULATIONS
Text pages 198–223

This chart is included for no other purpose than to demonstrate the complexity of the skull.

		CRANIAL BONES						FACIAL BONES							
		SPHENOID	ETHMOID	FRONTAL	*PARIETAL	*TEMPORAL	OCCIPITAL	MANDIBLE	*MAXILLA	*NASAL	*PALATINE	*LACRIMAL	*ZYGOMATIC	*INFER. NAS. CON	VOMER
CRANIAL BONES	SPHENOID		✔	✔	✔	✔	✔		✔		✔		✔		✔
	ETHMOID	✔	✔	✔					✔	✔	✔	✔		✔	✔
	FRONTAL	✔	✔		✔				✔	✔		✔	✔		
	*PARIETAL	✔		✔	✔	✔	✔								
	*TEMPORAL	✔			✔		✔	✔					✔		
	OCCIPITAL	✔			✔	✔									
FACIAL BONES	MANDIBLE					✔									
	*MAXILLA	✔	✔	✔					✔	✔	✔	✔	✔	✔	✔
	*NASAL		✔	✔					✔	✔					
	*PALATINE	✔	✔						✔		✔			✔	✔
	*LACRIMAL		✔	✔					✔					✔	
	*ZYGOMATIC	✔		✔		✔			✔						
	*INFER. NAS. CON		✔						✔		✔	✔			
	VOMER	✔	✔						✔		✔				

(Chart may be read from either direction)
* means paired bones

Quote: "Except for the mandible and the ossicles of the middle ear, the bones of the adult skull are joined by rigid sutures or synchondroses. In effect, the cranium of the mature adult is essentially a single complex bone."

Keith L. Moore, 1985

Quote: "Doctors use the fontanelles to gain information about the baby's brain and neurological development. For example, the crying infant will have a pulsating and slightly bulging soft spot that resolves once the baby is quiet. Bulging soft spots in the absence of the healthy infantile wailing can, however, be a sign your pediatrician needs to know about and act upon immediately. A variety of serious conditions can be heralded by this sign, ranging from meningitis to tumors. Conversely, if a baby's soft spot becomes sunken or depressed, it can be a sign of dehydration (loss of water)."

H. Markel and F. Oski, 1996

FIGURE 4.4 LEFT
MAXILLA AND PALATINE
BONES AS SEEN FROM
BENEATH. (from Zemlin
and Stolpe)

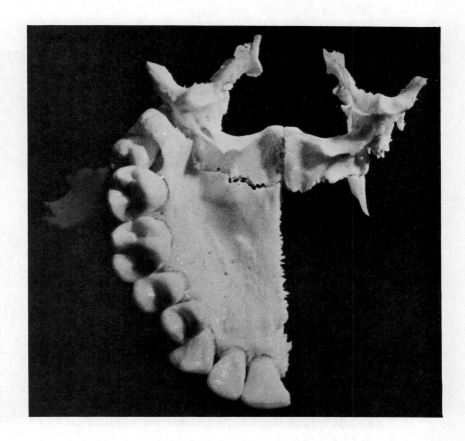

FIGURE 4.5 THE ARTICULATED SPHENOID, VOMER, AND PALATINE BONES AS SEEN FROM
THE FRONT. (from Zemlin and Stolpe)

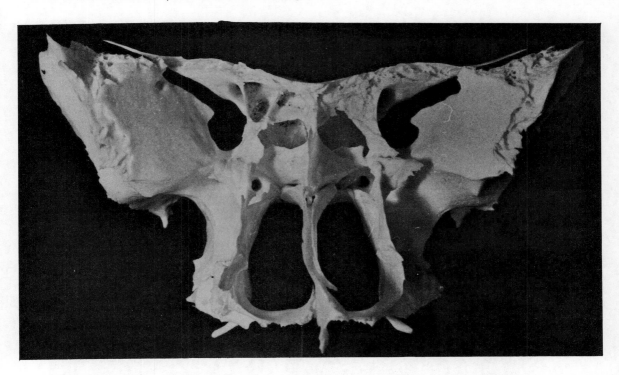

FIGURE 4.6 THE MAXILLAE, ARTICULATED WITH THE ETHMOID AND RIGHT ZYGOMATIC
BONES. (from Zemlin and Stolpe)

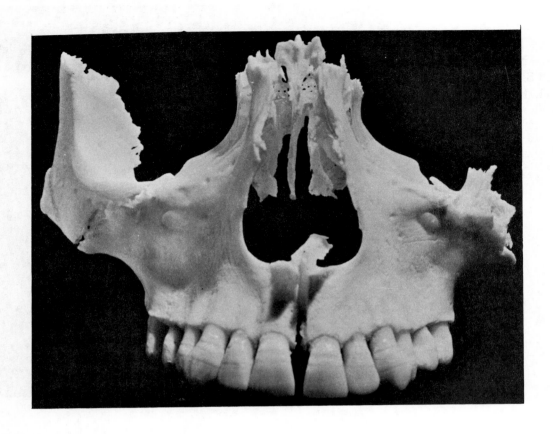

FIGURE 4.7 LEFT MAXILLA, ARTICULATED WITH PALATINE, ETHMOID AND SPHENOID
BONES, AS SEEN FROM THE FRONT. (from Zemlin and Stolpe)

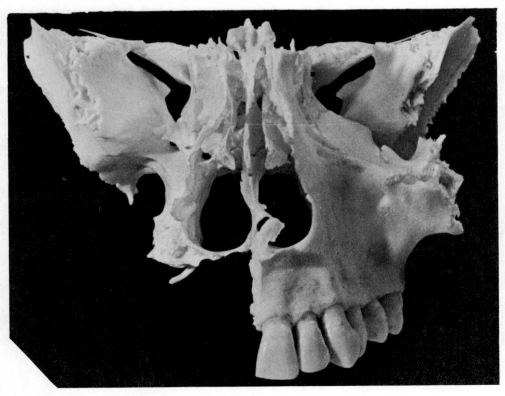

No. 4-9 THE PARANASAL SINUSES AND THE MASTOID AIR CELLS
Text pages 223–225

Description	Frontal	Maxillary	Ethmoid	Sphenoid
paranasal sinuses	✔	✔	✔	✔
present at birth		✔	✔	
located behind the superciliary arch	✔			
largest sinuses		✔		
drain into the nasal cavity	✔	✔	✔	✔
because of their complexity, often called labyrinth			✔	
mucous membrane continuous with that of the nasal cavity	✔	✔	✔	✔
an infection in this sinus may produce pain similar to that of a dental abscess		✔		
reduce the weight of the skull	✔	✔	✔	✔

1. How do the sinuses affect speech production?   1. _____

_____   _____

2. Are the mastoid cells true sinuses?   2. _____

3. The mucous membrane lining the mastoid air cells is   3. _____
    continuous with the lining of the _____ cavity.

Question: Why is it particularly important to control and contain infection of the ethmoid
            and frontal sinuses and the mastoid air cells?

No. 4-10 THE CAVITIES OF THE VOCAL TRACT
Text pages 226–228

1. The cavities bounded by the lips and cheeks externally and   1. _____
    the teeth and gums internally are the _____ cavities.

2. The palatoglossal arch forms the posterior boundary of the   2. _____
    _____ cavity.

3. The palatoglossal and palatopharyngeal arches form a   3. _____
    channel of communication between the _____ and
    _____ cavities.   _____

4. Inferiorly, the pharyngeal cavity is continuous with the _____.   4. _____

5. List four cavities with which the pharynx communicates.   5. _____

_____

_____

_____

6. The soft palate controls communication between the   6. _____
    _____ and the _____.

7. The level of the hyoid bone is the boundary between the   7. _____
    _____ and the _____.   _____

8. The pharyngeal cavity extends from the base of the _____   8. _____
    to the level of the *2ⁿᵈ / 4ᵗʰ / 6ᵗʰ* cervical vertebra posteriorly
    and the _____ cartilage anteriorly.   _____

_____

No. 4-11 THE NOSE AND NASAL CAVITIES, MUSCLES OF THE NOSE (VOCABULARY)
Text pages 228–231

alar	nares	procerus
apex	nasal dilators	quadratus labii superior
bridge	nasal meatuses	(angular head)
ciliary action	nasal turbinates (chonchae)	respiratory
choanae	nasalis	root
depressor septi	olfactory	septal
maxillae	palatine bones	

1. paired muscles that may narrow or constrict nostrils (2)

   1. _____

     _____

2. only part of the nose that has a bony framework

   2. _____

3. form the floor of the nasal cavity (2)

   3. _____

4. located at junction of nasal bones and frontal bone

   4. _____

5. the tip of the nose

   5. _____

6. nostrils

   6. _____

7. cartilage that forms the tip of the nose

   7. _____

8. tract that begins in the lower part of the nose

   8. _____

9. organ situated in upper part of nose

   9. _____

10. paired muscles which may enlarge the nostrils (2)

   10. _____

      _____

11. muscle inserting in the area between the eyebrows; contraction may help produce a frown

   11. _____

12. provide communication between nasal cavities and the nasopharynx

   12. _____

13. comprise the lateral walls of the nasal cavity (2)

   13. _____

      _____

14. propels mucus toward the pharynx

   14. _____

15. Cartilage that completes separation of nostrils at the base of the nose.

   15. _____

Note: *Rhin-, rhino-* pertain to nose, as in rhinoceros. Speculate on the meanings of the following terms:

    rhinitis
    rhinolalia (Gr. *lalia*, speech)
    rhinoplasty (Gr. *plassein*, to form)
    rhinoscopy

1. Which articulator can move most rapidly?

2. What accounts for the reddish hue of the vermilion zone of the lips?

3. The inner surface of the lip is also called the _____ surface.

4. The frenulum of the lower lip connects the lip with the _____ process of the _____ .

5. The frenulum of the upper lip connects the lip with the _____ process of the _____ .

6. The term labial pertains to _____ .

7. The buccal fat pad, which is more important in *infants / adults*, is sometimes called the _____ pad.

8. The parotid, submaxillary, and sublingual glands secrete _____ .

9. Modification of the resonant characteristics of the oral cavity is important for the production of *vowels / consonants*.

10. Constriction of the articulators is particularly important for the production of *vowels / consonants*.

1. _____

2. _____
   _____

3. _____

4. _____
   _____

5. _____
   _____

6. _____

7. _____
   _____

8. _____

9. _____

10. _____

Quote: "Children who persistently breathe through their mouths as a result of nasal obstruction may develop what is known as the 'long-face syndrome' or what was previously termed 'adenoid facies.' There are probably genetic factors that determine which children with nasal obstruction will develop this syndrome. There is also a small group of children who develop the syndrome in the absence of nasal obstruction.

Nasally obstructed children who develop the 'long-face syndrome' tend to adopt an extended head position, with a lowered tongue or a depressed mandible posture, in order to establish a better oral airway. The nasal obstruction may have a variety of causes, including mucosal swelling from allergy, deviation of the nasal septum, enlarged turbinates, or enlargement of the adenoid pad. In other words, the 'adenoids facies' may be caused by any form of nasal obstruction.

The head extension and downward mandibular posture may produce a narrowing of the dentition, and these nasally obstructed individuals are more likely to exhibit a class II malocclusion. As the problem progresses there can be a retrognathic mandible, a high palatal vault, a short upper lip, and flabbiness of the perioral muscles. Moreover, in the established condition, habitual mouth breathing may persist even after the underlying nasal obstruction has been cleared.

Orthodontic referral is necessary for children who have abnormalities in the position or alignment of their teeth. It is important to recognize the potential orthodontic problems and abnormalities of facial growth that may result from untreated nasal obstruction in early childhood. Any child with persistent nasal obstruction should be referred promptly to an otolaryngologist."

B. Bingham, M. Hawke, and P. Kwok, 1992

No. 4-13 MUSCLES OF THE FACE AND MOUTH
Text pages 233–239

In this exercise you are asked to demonstrate the action of a particular muscle or group of muscles, but remember these muscles are not acting alone. As stated in the text, ". . . the muscles of the face and lips are so intrinsically related they exhibit functional unity."

Before demonstrating the action of a muscle, identify it in one of the illustrations of the facial musculature. As you're making the suggested "grimaces" can you feel any changes occurring in the area of the muscle?

Orbicularis Oris Muscle (L. *circular mouth*)	Close your mouth.	Pucker your lips.

TRANSVERSE FACIAL MUSCLES

buccinator (L. *bugler*)	Retract the corners of your mouth without smiling or frowning.	Fill your cheeks with air, then gradually expel it.
risorius	Retract the corners of your mouth without smiling or frowning.	Fill your cheeks with air, then gradually expel it.

ANGULAR FACIAL MUSCLES

levator labii superior	Elevate your upper lip.	Slightly evert your upper lip.
levator labii superior alaeque nasi	Elevate your upper lip.	Dilate your nostrils.
zygomatic minor	Elevate your upper lip.	Attempt to deepen the furrows between your nose and the corners of your mouth.
zygomatic major	Smile broadly.	
depressor labii inferior	As you chew, notice that your lower lip is drawn downward and slightly lateralward.	

VERTICAL FACIAL MUSCLES

mentalis	As you drink from a glass, notice that your lower lip is everted.
depressor anguli oris	When your mouth is wide open, notice that the angles of the mouth descend; then bring your upper lip down to meet your lower lip.
levator anguli oris	Attempt to deepen furrows as for zygomatic minor (above).

PARALLEL FACIAL MUSCLES

incisivis labii superior	Pucker your lips.
incisivis labii inferior	Do the impossible! Try to pucker just your lower lip.
Platysma Muscle	Jut out your jaw as far as you can; then retract the corners of your mouth.

It is important that you understand how the buccinator muscle and the posterior pharyngeal constrictor attach to the pterygomandibular raphe or ligament.

Question: Do you think this old adage true? "It takes more muscles to frown that it does to smile."

FIGURE 4.9 SUPERFICIAL FACIAL MUSCULATURE (MUSCLES OF FACIAL EXPRESSION) AS SEEN FROM THE FRONT.

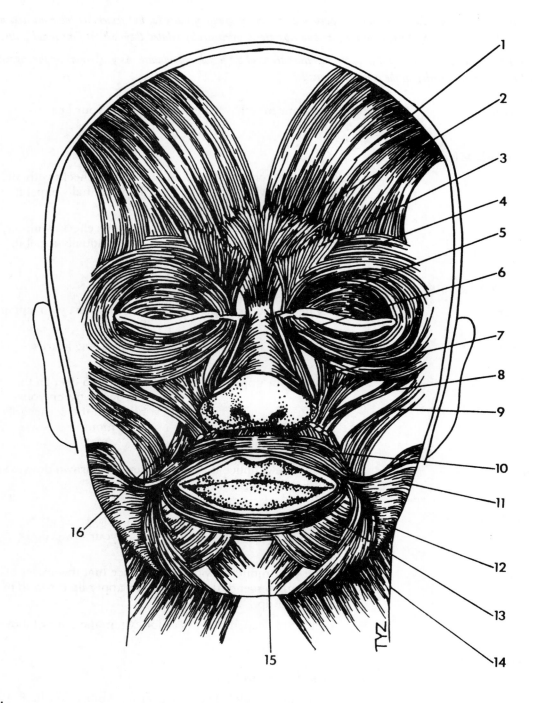

Identify:

1. epicranius pars frontalis
2. procerus
3. corrugator
4. orbicularis oculi
5. nasalis
6. levator labii alaeque nasi
7. levator labii superioris
8. zygomatic minor

9. zygomatic major
10. orbicularis oris
11. risorius
12. depressor anguli oris
13. depressor labii inferioris
14. platysma
15. mentalis
16. levator anguli oris

FIGURE 4.10 SUPERFICIAL FACIAL MUSCULATURE (MUSCLES OF FACIAL EXPRESSION), MUSCLES OF THE SCALP AND NECK AS SEEN IN PERSPECTIVE FROM THE SIDE.

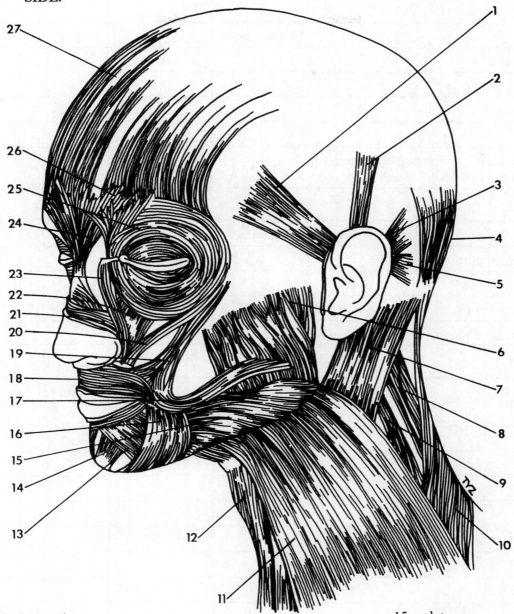

Identify:

1. auricularis anterior
2. auricularis superior
3. auricularis posterior
4. epicranius pars occipitalis
5. auriculus posterior
6. masseter
7. sternocleidomastoid
8. splenius
9. levator scapulae
10. trapezius
11. platysma
12. sternohyoid
13. depressor labii inferioris
14. mentalis

15. platysma
16. depressor anguli oris
17. risorius
18. orbicularis oris
19. zygomatic major
20. zygomatic minor
21. nasalis
22. levator labii superioris
23. levator labii alaeque nasi
24. procerus
25. orbicularis oculi
26. corrugator
27. epicranius pars frontalis

Description	Part of Tooth:	Crown	Root	Neck
covered by enamel		✔		
covered by cementum			✔	
transitional area				✔
largest part			✔	
in children, the anatomical portion of a permanent tooth which is larger than the functional portion		✔		

Description	Tissues:	Enamel	Cementum	Dentin	Dental Pulp
predominant solid portion of tooth				✔	
inner tissue, rich in nerves and blood tissues					✔
hardest substance in the body		✔			
bonylike substance covering root			✔		

Description	Types of Teeth:	Incisors	Canines	Premolars	Molars
chisel-shaped		✔			
single root		✔	✔	✔	
pointed, tusklike			✔		
eye teeth			✔		
cuspids			✔		
bicuspids				✔	
suited for crushing and grinding				✔	✔
suited for biting and shearing		✔			
suited for ripping and tearing			✔		
flat, broad surfaced					✔
have more than one root				✔	✔
largest teeth					✔

Equivalent Terms	Molars:	First	Second	Third
wisdom teeth				✔
six-year molars		✔		
twelve-year molars			✔	

No. 4-15 THE LIFE CYCLE OF A TOOTH; THE DENTAL ARCHES
Text pages 241–247

Teeth in Half a Dental Arch Arches:	Permanent	Deciduous
one central incisor	✔	✔
one lateral incisor	✔	✔
one cuspid (canine)	✔	✔
two bicuspids (premolars)	✔	
first molar	✔	✔
second molar	✔	✔
third molar	✔	

1. What are the four periods in the life cycle of a deciduous or permanent tooth?

    1. _____
    _____
    _____
    _____

2. Large multinucleated cells which resorb bony tissue, thus making way for an erupting tooth, are called _____.
Bone cells are called _____.
Bone-forming cells are called _____.

    2. _____
    _____
    _____

3. When do teeth stop erupting.

    3. _____

4. When does attrition of a tooth begin?

    4. _____

5. Most children have all their deciduous teeth by the time they are _____ years old.

    5. _____

6. The first permanent molars erupt immediately behind the *first / second* deciduous molars.

    6. _____

7. The deciduous molars are replaced by the permanent _____.

    7. _____

8. The *superadded* permanent teeth, those having no deciduous predecessors, are the _____.

    8. _____

Questions: Cartoons depicting the pulling of a deciduous tooth usually feature a huge molar, complete with vicious-looking roots, dangling from a string. How would you explain the shedding of baby teeth to a child?

A parent asks, "Why should I spend money on baby teeth when they're just going to fall out anyway?" How would you respond?

FIGURE 4.11 FULLY ERUPTED DECIDUOUS DENTITION.
Take note of the overbite and overjet.

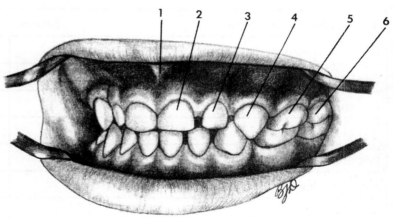

Identify:

1. labial frenum
2. upper central incisor
3. upper lateral incisor
4. upper canine
5. upper first molar
6. upper second molar

FIGURE 4.12 PERMANENT DENTITION.

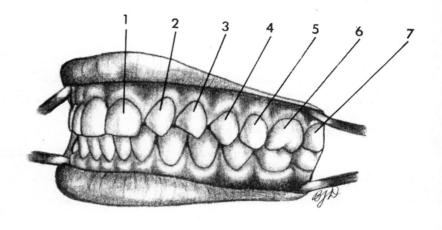

Identify:

1. central incisor
2. lateral incisor
3. canine
4. first premolar
5. second premolar
6. first molar
7. second molar

Note: The third molars are not shown in this illustration because the model's third molars had not yet erupted.

Question: About how old might she have been?

FIGURE 4.13 THE SURFACES OF THE DENTITION.

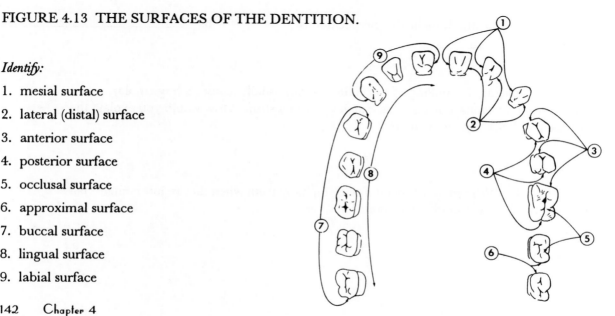

Identify:

1. mesial surface
2. lateral (distal) surface
3. anterior surface
4. posterior surface
5. occlusal surface
6. approximal surface
7. buccal surface
8. lingual surface
9. labial surface

1. Which arch is usually larger in diameter, the mandibular or maxillary?

1. _____

2. A space between two teeth, most frequently the upper central incisors, is called a _____.

2. _____

3. If you are unable to approximate your front teeth you have a/an _____ bite. If your molars don't meet you have a/an _____ bite.

3. _____
   _____

Description	Angle's Class	I	II	III
increased overjet			✔	
increased facial height				✔
normal profile		✔		
prognathic mandible				✔
normal jaw relationships; abnormal positioning of teeth		✔		
retruded mandible			✔	

Define and/or sketch for your own purposes the following tooth positions:

OCCLUSAL RELATIONSHIP OF FIRST MOLARS

infraversion   supraversion

Normal (Class I if teeth are malpositioned)

labioversion   linguaversion

Class II

distoversion   mesioversion

axiversion   torsiversion   Class III

Note: Hypertrophy (overgrowth) of gum tissue is a relatively common side effect of Dilantin, medication used to control seizures. Good dental hygiene may slow the process, but surgical removal of excess tissue is sometimes necessary.

Quote: (referring to mandibular asymmetry resulting from unilateral posterior crossbite associated with a lateral shifting of the mandible) "In the crossbite relationship the maxillary buccal cusps occlude in the central groove of the lower posterior teeth. . . . expansion of the maxillary dentition by orthodontic means will correct the problem . . . this type of crossbite must be recognized and corrected early."

L. T. Swanson and J. E. Murray, 1978 (Whitaker and Randall)

FIGURE 4.14 THE ORAL CAVITY AND SOME ADJACENT STRUCTURES.

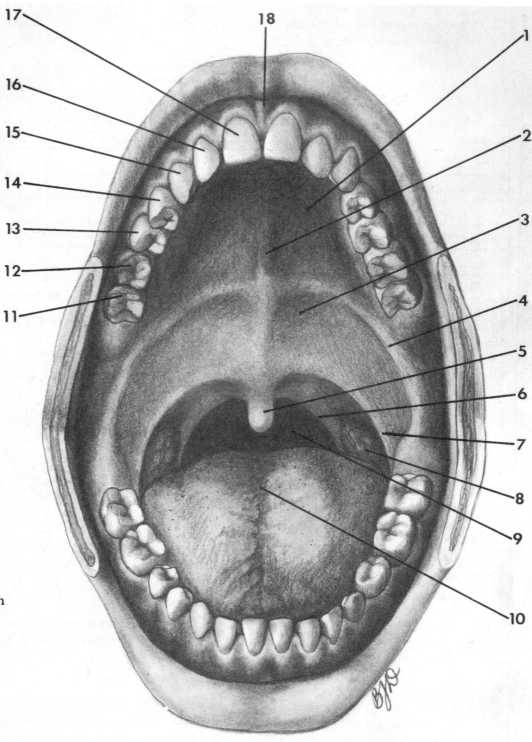

Identify:

1. hard palate
2. median raphe
3. soft palate
4. pterygomandibular raphe
5. uvula
6. palatoglossal arch
7. palatopharyngeal arch
8. palatine tonsil
9. pharyngeal wall
10. median longitudinal sulcus
11. second molar
12. first molar
13. second premolar
14. first premolar
15. canine
16. lateral incisor
17. central incisor
18. labial frenum

1. What are the primary biological functions of the tongue? (3)

1. _____

   _____

   _____

2. Why is the tongue capable of assuming so many different positions with such rapidity and subtlety?

2. _____

   _____

3. What are the two anatomical divisions of the tongue?

3. _____

   _____

4. Why, in human beings, does the upper airway turn almost 90° in a forward direction? How does this affect tongue movement?

4. _____

   _____

5. The characteristic roughness of the anterior two-thirds of the tongue is due to the presence of _____, while the smoother posterior third contains _____ glands and _____ glands.

5. _____

   _____

   _____

   _____

6. The lymph follicles on the root of the tongue comprise the _____ tonsil.

6. _____

7. The "skeleton" of the tongue is composed of a dense felt-like network of fibrous connective tissue called the _____.

7. _____

8. Do the extrinsic and intrinsic muscles follow separate pathways in the tongue?

8. _____

Description	Functional Divisions:	Tip	Blade	Front	Back
just below the hard palate				✔	
nearest the incisors		✔			
beneath the soft palate					✔
just below the maxillary alveolar ridge			✔		

Structures	Attachments of the Tongue Musculature:	Superior	Inferior
pharynx			✔
palate		✔	
hyoid bone			✔
epiglottis (by ligaments)			✔
base of the skull		✔	
mental symphysis (inner surface)			✔

No. 4-18 THE INTRINSIC MUSCLES OF THE TONGUE
Text pages 253-254 *Study figures in text.*

Action or Description	Sup. Long.	Inf. Long.	Transverse	Vertical
largest intrinsic muscle	✔			
courses between genioglossus and hyoglossus		✔		
may flatten the tongue				✔
may groove the tongue	✔			
may narrow and elongate the tongue			✔	
may shorten the tongue	✔	✔		
may pull tongue tip up	✔			
may pull tongue tip down		✔		

No. 4-19 THE EXTRINSIC MUSCLES OF THE TONGUE
Text pages 254-256

Action or Description -glossus:	Genio-	Stylo-	Palato-	Hyo-
borders oropharyngeal isthmus			✔	
strongest and largest extrinsic muscle	✔			
a bundle of its fibers is sometimes called the *chondroglossus*				✔
originates on the temporal bone		✔		
originates on the posterior surface of the mandibular symphysis	✔			
antagonist of the genioglossus		✔		
lower fibers insert on the hyoid bone	✔			
may groove the dorsum of the tongue	✔	✔	✔	
may protrude the tongue	✔			
may retract the tongue	✔			✔
may elevate the hyoid bone				✔

Questions: How does tongue movement differ from simpler muscle movement?

Could the inability to produce a certain sound or group of sounds correctly be the only symptom of minor anatomical and/or physiological deviation?

What enables some people with severe deformities of the speech mechanism to produce normal or relatively normal speech?

How might the following diagnoses affect your clinical approach to a speech or language problem? organic, functional, idiopathic (of unknown causation)

FIGURE 4.15 EXTRINSIC TONGUE MUSCULATURE AND SOME ASSOCIATED STRUCTURES.

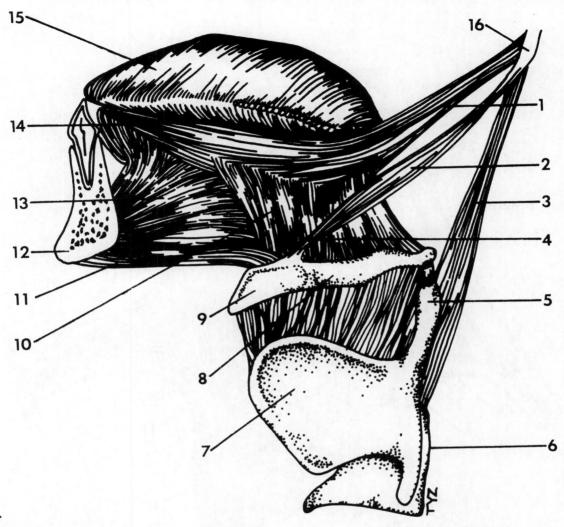

Identify:

1. styloglossus m.
2. stylohyoid m.
3. stylopharyngeus m.
4. lesser horn of hyoid bone
5. superior horn of thyroid cartilage
6. cricoid cartilage
7. thyroid lamina
8. greater horn of hyoid bone

9. body of hyoid bone
10. hyoglossus m.
11. geniohyoid m.
12. mandible (cut)
13. genioglossus m.
14. inferior longitudinal m.
15. dorsum of tongue
16. styloid process of temporal bone

When conducting an orofacial examination you will usually observe the execution of various tongue movements, e.g., protruding, pointing, flattening, curling edges, curling tip, grooving, sweeping lips. You might spend a few minutes watching each other as you perform or attempt to perform these movements.

Questions: In which types of movements do you see the greatest variability from person to person?

Could the way you learn to produce certain sounds be influenced by individual differences in tongue mobility, or conversely, could tongue mobility be influenced by the way you learn to produce sounds?

Might the way you articulate certain sounds, e.g., (s) (z) affect your clinical techniques for stimulating correct production of these sounds.

FIGURE 4.16 TOP VIEW OF LARYNX, TONGUE, AND ASSOCIATED STRUCTURES.

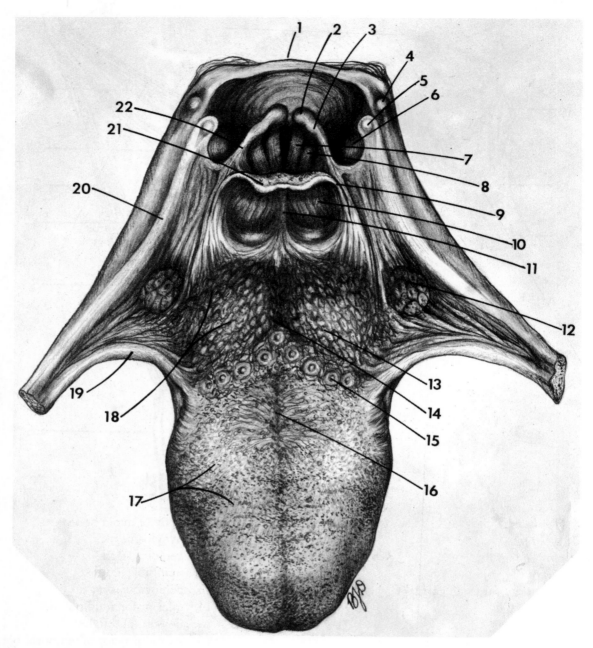

Identify:

1. posterior pharyngeal wall
2. highlight of corniculate cartilage
3. highlight of cuneiform cartilage
4. tip of greater horn of hyoid
5. tip of superior horn of thyroid
6. pyriform sinus
7. true vocal fold
8. false vocal fold
9. lateral glossoepiglottic fold
10. vallecula
11. middle glossoepiglottic fold
12. palatine tonsil
13. lingual tonsil (base of tongue)
14. foramen cecum
15. vallate (circumvallate) papilla
16. longitudinal sulcus
17. blade of tongue
18. root of tongue
19. palatoglossal arch
20. palatopharyngeal arch
21. epiglottis
22. aryepiglottic fold

1. Does the *extent* or *rapidity* of mandibular movement seem to be of much greater importance in speech production?

2. How do the mandible and the tip of the tongue compare in terms of maximum rate of movement?

3. Disturbances of the temporomandibular joint may be the *cause / result / result or cause* of malocclusion.

4. Does the temporomandibular ligament *restrict* or *enhance* the movement of the mandible?

5. What term describes a partial dislocation of the jaw in which the condyle slips under the zygomatic process of the temporal bone?

6. Why are the left and right temporomandibular joints considered a single bilateral articulation rather than two unilateral articulations?

1. _____

2. _____

3. _____

4. _____

5. _____

6. _____

Description/Action TM Joint:	Upper	Lower
between articular disc and condyle		✔
between articular disc and articular eminence of mandibular fossa	✔	
depression and elevation of mandible		✔
protraction and retraction of mandible	✔	
lateral movement of the mandible	✔	
performs rotational, hingelike movement		✔
performs translatory, gliding movement	✔	

To discover how well you can speak without mandibular movement, talk while holding a pen or pencil between your teeth as though it were a cigarette holder.

Question: As you observe "normal" speakers do you ever notice a slight lateral excursion of the mandible on (s) (z) (sh) (ch) (zh)?

Note:

TEMPOROMANDIBULAR DYSFUNCTION

Symptoms:
1. pain, sometimes referred to the ear
2. cracking or grating sounds during jaw movement
3. limited jaw opening

Causes:
1. dislocation
   a. acute (due to trauma)
   b. chronic (result of muscular imbalance)
2. arthritis
3. ankylosis (fusion of the joint)

Effects:
1. unilateral ankylosis during childhood may result in mandibular and facial asymmetry
2. bilateral ankylosis during childhood may result in a "bird-face" (small receding chin and malocclusion)

digastric   medial pterygoid
geniohyoid   mylohyoid
lateral pterygoid   sternohyoid
masseter   temporalis

1. depress the mandible (4)
   (check those which are also laryngeal elevators)

   1. _____

     _____

     _____

     _____

2. elevate the mandible (3)

   2. _____

     _____

     _____

3. retract the mandible (2)

   3. _____

     _____

4. form mandibular sling, functionally uniting the upper and lower jaw (2)

   4. _____

     _____

5. fixes the hyoid bone when the mandible is opened against resistance

   5. _____

6. sheet of muscle fiber forming the floor of the mouth

   6. _____

7. fan-shaped; arises on temporal fossa and inserts on coronoid process of ramus of mandible

   7. _____

8. anterior and posterior bellies coursing downward from mandible and mastoid processes respectively, meet at a tendon which attaches to the hyoid bone

   8. _____

9. bilateral contraction may protract* mandible, while unilateral contraction causes lateral or grinding movements

   9. _____

10. slowly contracting, powerful; exerts force in area of molars

   10. _____

11. cylindrical; just superior and lateral to midline raphe of mylohyoid muscles

   11. _____

12. thick, quadrilateral; arising from pterygoid fossa and lateral plate, maxilla, and perpendicular plate of palatine bone and inserting on medial surface of ramus and angle of mandible

   12. _____

13. arises from greater wing and pterygoid plate of sphenoid, then converges to insert on condyle of mandible

   13. _____

14. thick, flat, quadrilateral; arises on zygomatic arch and inserts on lateral surface of ramus and angle of mandible

   14. _____

*extend or protrude

No. 4-22 THE PALATE
Text pages 264–265

Description	Maxillae (palatine processes)	Palatine Bones (horizontal plates)
posterior borders are free		✔
anterior three-fourths of hard palate	✔	
form posterior nasal spine		✔
site of rugae (membranous, not bony)	✔	
membranous midline raphe	✔	✔
posterior one-fourth of hard palate		✔

1. What is the role of the palate in speech production?

1. _____
_____

2. What is a torus palatinus? What is its incidence?

2. _____
_____

3. As you watch people grin, you may be able to predict the general contours of their palatal vaults. Explain why.

3. _____
_____

4. The posterior margin and the midline are the *thickest / thinnest* portions of the bony palate.

4. _____

5. What is another name for the soft palate?

5. _____

6. What attaches the soft palate to the palatine bones and also forms the fibrous skeleton of the soft palate?

6. _____

7. For which speech sounds is the soft palate lowered?

7. _____

8. What is the position of the soft palate during normal breathing?

8. _____

In the course of an orofacial examination you will describe palatal vault, tongue size, and other factors requiring judgment rather than measurement. The best way to establish bases for these judgments is to scrutinize the mouths of many children and adults, particularly those having normal speech.

Question: How may the concepts of anatomical and physiological variability affect your diagnosis of speech and language disorders?

No. 4-23 THE MUSCLES OF THE PALATE
Text pages 265–270

Description/Action	Levator Palati	Musculus Uvulae	Palato-glossus	Palato-pharyngeus	Tensor Palati
depressor relaxors (2)			✔	✔	
depressor tensors					✔
palatal elevators (2)	✔	✔			
primarily responsible for velopharyngeal closure	✔				
shorten and lift soft palate		✔			
raise back of tongue or lower soft palate			✔		
form inner structure of anterior arch			✔		
form inner structure of posterior arch				✔	
form muscular sling for soft palate	✔				
contraction opens auditory (Eustachian) tubes					✔
form bulk of soft palate	✔				
sphincterlike action (2)			✔	✔	
arise from petrous portion of temporal bone and posteromedial plate of auditory tube	✔				
arise from pterygoid plate and anterolateral wall of the auditory tube, then wind around pterygoid hamulus					✔
arise from soft palate, pterygoid hamulus, and auditory tube				✔	
arise from nasal spines and palatal aponeurosis		✔			
divide each palatopharyngeus muscle in two	✔				
primarily muscles of deglutition				✔	
arise from anterior surface of palatine aponeurosis and insert on sides of the tongue			✔		

Developing a visual construct of the palatal musculature and an appreciation of its anatomical and physiological complexity is an important prerequisite to the study of cleft palate and velar insufficiency. Illustrations may be helpful, but tracing the pathways of the muscles on an actual skull will be more meaningful.

Questions: As you say the following sentence, can you feel the movement of your soft palate as you say the nasal sounds [m] and [n]? My mother's name is Nancy. Is it easier to feel the movement of the other articulators? Why might it be difficult to modify palatal movement?

FIGURE 4.17 SCHEMATIC SAGITTAL SECTION OF HEAD SHOWING TONGUE MUSCULATURE, SOFT PALATE, AND SUPERIOR PHARYNGEAL CONSTRICTOR REGION.

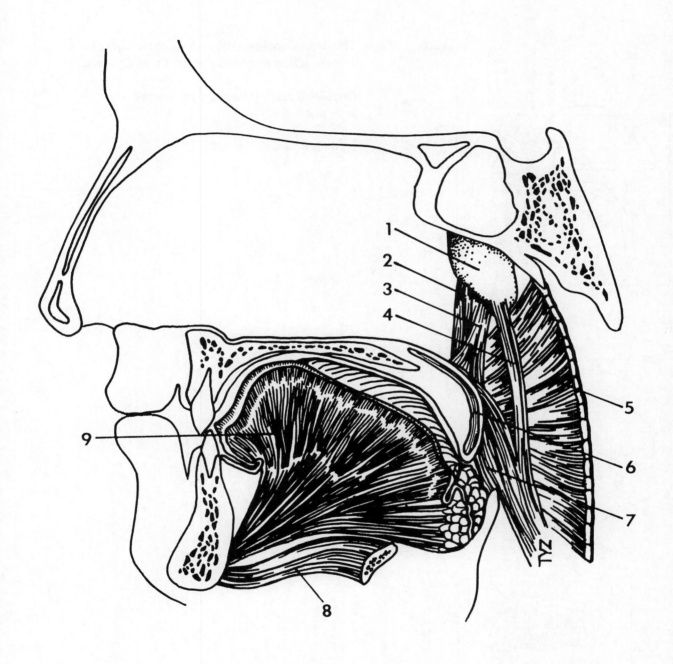

Identify: 1. cartilage of auditory (Eustachian) tube

2. tensor palati m.

3. levator palati m.

4. salpingopharyngeus m.

5. superior pharyngeal constrictor

6. musculus uvulae

7. palatopharyngeus m.

8. geniohyoid m.

9. genioglossus m.

See Fig. 3.7, page 97, for an illustration of the inferior pharyngeal constrictor.

FIGURE 4.18 SCHEMATIC REPRESENTATION OF THE VELOPHARYNGEAL MUSCLES.

From Fritzell, *The Velopharyngeal Muscles in Speech*, 1969, by permission.

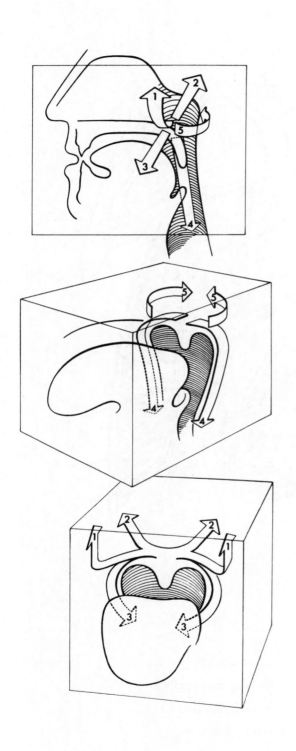

The arrows indicate the approximate direction of their action and influence on the soft palate.

Describe the action of the muscles represented in the illustration:

1. tensor palati

2. levator palati

3. palatoglossus

4. palatopharyngeus

5. superior pharyngeal constrictor

No. 4-24 THE TONSILS
Text pages 270–273

Description	Palatine	Pharyngeal	Lingual
inferior portion of Waldeyer's ring			✔
superior portion of Waldeyer's ring		✔	
lateral portion of Waldeyer's ring	✔		
the adenoids		✔	
called "the tonsils"	✔		
cover the root of the tongue			✔
can be seen in oral exam	✔		
on the posterior wall of the nasopharynx		✔	
may contribute to velopharyngeal closure		✔	
chronic enlargement may result in mouth breathing		✔	
may help develop immunity to bacterial infections	✔	✔	✔
chronic hypertrophy may affect the auditory tube		✔	
begin to atrophy about the age of puberty	✔		
hit growth peak at 9–10 years		✔	

1. What effect will enlarged palatine tonsils have on speech?   1. _____

_____

Quote: "As recently as the 1960s, it was rare to find a child older than five who still had his tonsils or adenoids. At one point or another, these small pieces of lymphoid tissue have been named as the cause of everything from ear infections to bedwetting to tuberculosis to hyperactivity."

H. Markel and F. Oski, 1996

Note: Infectious criteria for tonsillectomy have not been established, but four to five episodes of tonsillitis per year for a period of two years is mentioned in the literature. Other indications for surgery may be obstruction of the airway and oral/facial problems. Because of the generous blood supply in the area of the palatine tonsils, there is a very small, but significant chance of hemorrhage following tonsillectomy.

Adenoidectomy is no longer the constant companion of tonsillectomy. Speech/language pathologists are sometimes asked to help determine if removal of the adenoids might reveal or aggravate velopharyngeal insufficiency.

Question: Speech-language pathologists and audiologists often refer children to physicians to determine probable causes of mouth breathing, hearing loss, and disturbances of facial growth. What are your responsibilities in the referral process?

No. 4-25 THE PHARYNX
Text pages 273–274

1. The pharynx forms the upper part of what systems? 1. ____________ ____________

2. The length of the pharynx, from its inferior border to the base of the skull, is approximately *3½ / 4¾ / 5¾* inches, while its extreme width superiorly is about *1⅗ / 2⅘ / 3⅗* inches. 2. ____________ ____________

3. How does the muscle arrangement of the pharynx resemble that of the gut? 3. ____________ ____________

4. In terms of speech production, the pharynx is primarily a/an ____________. 4. ____________

5. Is modification of the size and configuration of the vocal tract greatly influenced by movement of the pharyngeal walls? 5. ____________

Description	Nasopharynx	Oropharynx	Laryngopharynx
divisions of the pharynx	✔	✔	✔
extends from the soft palate superiorly to the level of the hyoid bone inferiorly		✔	
extends from the base of the skull superiorly, to the level of the soft palate inferiorly	✔		
extends from the level of the hyoid bone superiorly to the level of the sixth cervical vertebra inferiorly			✔
communicates with the nasal cavities and the auditory (Eustachian) tubes	✔		
communicates with the esophagus inferiorly and the aditus laryngis anteriorly			✔
communicates with the oral cavity		✔	
funnel-shaped			✔
widest point	✔		
most changeable portion		✔	

A more detailed description of the anatomy and physiology of the auditory (Eustachian) tube is given on pages 445–448 of the text.

Quote: A man with a wide pharynx and with a larynx that will resonate at a frequency between 2,500 and 3,000 hertz is likely to be able to develop a good singing voice more readily than a person who lacks these characteristics.

Johan Sundberg, 1977

FIGURE 4.19 THE PHARYNX AS SEEN IN PERSPECTIVE FROM BEHIND.

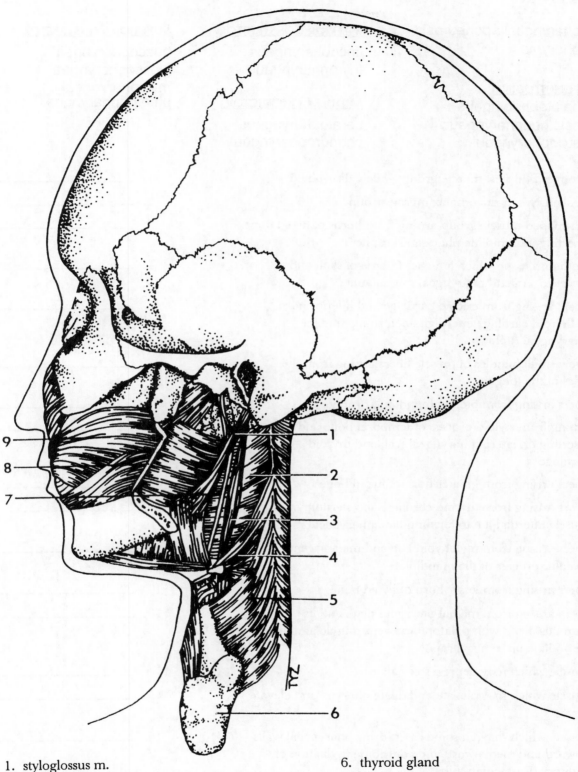

Identify: 1. styloglossus m.

2. superior pharyngeal constrictor

3. stylopharyngeus m.

4. middle pharyngeal constrictor

5. inferior pharyngeal constrictor

6. thyroid gland

7. stylohyoid ligament

8. pterygomandibular raphe

9. buccinator m.

pharyngeal aponeurosis
pharynx

INFERIOR CONSTRICTOR
cricopharyngeus
thyropharyngeus

SUPERIOR CONSTRICTOR
buccopharyngeal
glossopharyngeal
mylopharyngeal
pterygopharyngeal

LONGITUDINAL
palatopharyngeus
salpingopharyngeus
stylopharyngeus

MIDDLE CONSTRICTOR
ceratopharyngeus
chondropharyngeus

1. strongest, thickest muscle group; widely distributed

2. weakest, yet most complex muscle group

3. fan-shaped muscle group arising from horns of hyoid bone, inserting into midline pharyngeal raphe

4. tendinous tissue which attaches to base of skull and serves to suspend pharyngeal musculature

5. fibers arising from cricoid cartilage and inferior horn of thyroid cartilage interdigitate, forming midline pharyngeal raphe

6. a few small bundles of muscle fibers arising from the sides of the tongue

7. fibers arising from pterygomandibular raphe

8. arising from styloid process of temporal bone and inserting on lateral pharyngeal wall and thyroid cartilage

9. fibers arising from greater horn of hyoid bone

10. fibers arising from thyroid cartilage and sternothyroid muscle interdigitate to form midline pharyngeal raphe

11. fibers arising from mylohyoid line and the adjacent alveolar process of the mandible

12. fibers arising from lesser horn of hyoid bone

13. fibers arising from medial pterygoid plate and its hamulus blend with palatopharyngeus muscle and insert on midline pharyngeal raphe

14. muscle which may depress soft palate

15. muscle which may elevate and dilate pharynx and elevate larynx

16. muscle which may, in some cases, draw pharyngeal walls upward and medialward and contribute to dilation of auditory (Eustachian) tube

17. muscle which may contribute to sphincteric action of esophagus, functioning as pseudoglottis for esophageal speakers.

18. musculo-membranous tube suspended from base of skull with only loose attachments to surrounding structures

1. _____

2. _____

3. _____

4. _____

5. _____

6. _____

7. _____

8. _____

9. _____

10. _____

11. _____

12. _____

13. _____

14. _____

15. _____

16. _____

17. _____

18. _____

FIGURE 4.20 PHARYNX AND ASSOCIATED STRUCTURES AS SEEN FROM THE SIDE.

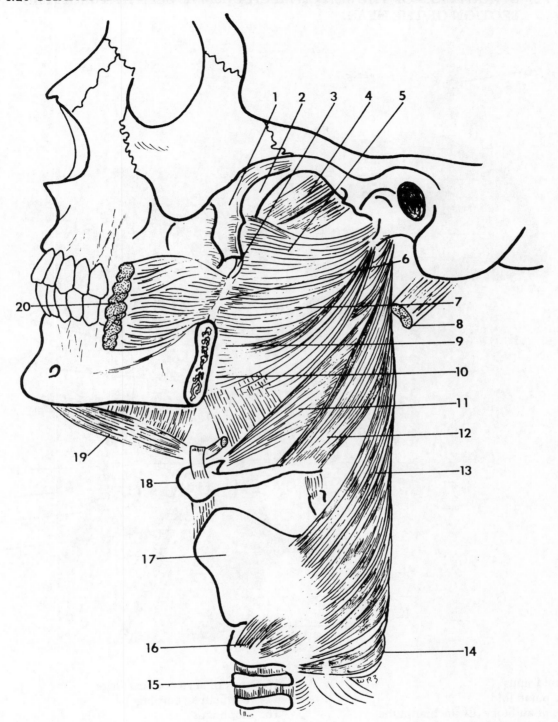

Identify:

1. lateral pterygoid plate
2. medial pterygoid plate
3. hamulus of medial pterygoid plate
4. lateral pterygoid m.
5. pterygopharyngeus m.
6. pterygomandibular ligament

7. mylopharyngeus m.
8. digastric (posterior belly)
9. buccopharyngeus m.
10. glossopharyngeus m.
11. chondropharyngeus m.
12. ceratopharyngeus m.
13. thyropharyngeus m.

14. cricopharyngeus m.
15. tracheal cartilage
16. cricothyroid m.
17. thyroid cartilage
18. hyoid bone
19. digastric (anterior belly)
20. buccinator m.

FIGURE 4.21 STRUCTURES OF THE SPEECH MECHANISM AS SEEN IN A SAGITTAL SECTION OF THE HEAD.

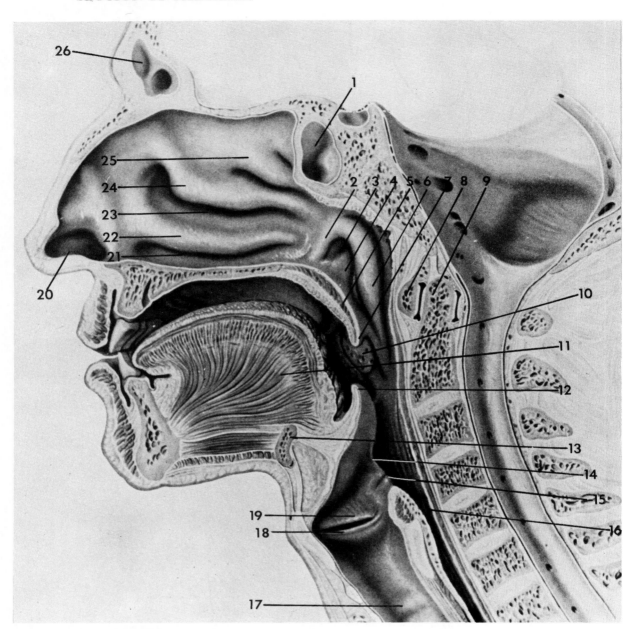

Identify:

1. sphenoid sinus
2. tensor palati fold
3. orifice of auditory (Eustachian) tube
4. levator palati fold
5. soft palate (musculus uvulae)
6. salpingopharyngeal fold
7. uvula
8. first cervical vertebra (atlas)
9. dens of second cervical vertebra
10. palatine tonsil
11. genioglossus m.
12. epiglottis
13. body of hyoid bone
14. aryepiglottic fold

15. apex of arytenoid cartilage (corniculate cartilage)
16. esophagus
17. trachea
18. true vocal fold
19. false vocal fold
20. nare
21. inferior nasal meatus
22. inferior concha (turbinate)
23. middle nasal meatus
24. middle concha (turbinate)
25. superior turbinate
26. frontal sinus

No. 4-27 THE VELOPHARYNGEAL MECHANISM
Text pages 278–280

1. The velopharyngeal mechanism modifies the coupling   1. _____
    between the _____ and the _____   _____
    cavities.

2. The degree of velopharyngeal closure is greater for *high* /   2. _____
    low vowels.

3. The degree of velopharyngeal closure is greatest for   3. _____
    vowels which are *isolated* / *adjacent to nasal consonants* /
    adjacent to nonnasal consonants.

4. Does Passavant's pad appear to play an important role in   4. _____
    velopharyngeal closure during speech production?

    Defend your answer. _____

    _____

5. Why might Passavant's pad occur more frequently in the   5. _____
    cleft palate population?

    _____

6. What pharyngeal movement appears to play an important role   6. _____
    in velopharyngeal closure?

7. Early scientists ascribed velopharyngeal closure to a   7. _____
    trapdoor-like movement. More current research attributes
    velopharyngeal closure to a _____ movement.

Question: Is it likely that the presence or absence of Passavant's pad could be determined
          in a routine oral examination of a person with adequate velopharyngeal closure?
          Defend your answer.

FIGURE 4.22 PHOTOGRAPH OF PASSAVANT'S PAD. (Soft palate had been partially excised.)

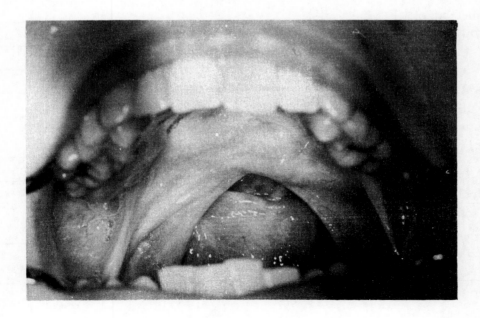

No. 4-28 SWALLOWING
Test pages 280–281

1. Another term for swallowing is _____.   1. _____

2. Impaired swallowing is called _____.   2. _____

Description Stage of Swallowing:	First	Second	Last
velopharyngeal seal		✔	
lips compressed; tongue pressing on hard palate	✔		
contraction of inferior constrictor			✔
sealing of laryngeal entrance		✔	
bolus enters esophagus; peristalsis begins			✔

Note: Speech-language pathologists are often involved in the development of feeding programs for severely handicapped children, and are becoming increasingly involved in the treatment of swallowing disorders.

No. 4-29 POSTNATAL GROWTH OF THE HEAD
Text pages 281–291

Prenatal growth of the facial region and palate is described in the text, pages 520–527. See also exercises 7-2 and 7-3, pages 279 and 281.

Description Growth of Skeleton:	Facial	Cranial
entirely dependent on growth of the brain		✔
dependent on muscular influences	✔	
dependent on growth of the teeth and tongue	✔	
ninety percent developed by age ten		✔
relatively large at birth		✔
relatively small at birth	✔	
grows faster after the first year	✔	

1. What are three methods of   1. _____
   studying skull growth?
   _____

   _____

2. The soft palate grows more rapidly during *infancy / puberty*.   2. _____

3. As growth proceeds, the soft palate becomes *more / less*   3. _____
   parallel to the posterior pharyngeal wall.

4. The growth of the facial skeleton is generally described   4. _____
   as progressing in a _____ and _____ direction.
   _____

5. $\dfrac{\text{horizontal depth of nasopharynx}}{\text{length of soft palate}}$ = approx. $\dfrac{x}{3}$ or _____%   5. x = _____ x ÷ 3 = _____ %

   An adenoidectomy may be contraindicated if the percentage is   _____
   much *lower / higher* than average.

   Why? _____

6. Why is mandibular height much greater in adults than in infants? 6. _____

7. What is a fontanelle? 7. _____

8. At what age does the bregmatic (anterior) fontanelle close? What might be the results of premature synostosis? 8. _____

9. What is the significance of the fronto-maxillary, zygomaticomaxillary, and zygomaticotemporal sutures? 9. _____

10. After the first year, the hard palate doubles in *length / width*. 10. _____

11. Growth in facial height coincides with eruption of the *incisors / molars*, while anteroposterior growth of the maxillae coincides with eruption of the _____. 11. _____

12. When is the angle of the ramus with the mandibular plane most obtuse? 12. _____

13. Resorption of alveolar bone may result <u>in</u> the loss of teeth/ <u>from</u> the loss of teeth / both. 13. _____

Description	Appositional	Interstitial
expansive growth due to cell multiplication		✔
growth due to deposition or formation at periphery	✔	
growth of bone	✔	
growth of muscle and cartilage		✔
growth of hyaline cartilage	✔	✔
primarily responsible for maxillary growth	✔	
primarily responsible for growth of mandibular condyle		✔
primarily responsible for increased height of the body of the mandible	✔	

Characteristics Skulls:	Infant	Adult
has unossified membranous regions	✔	
completely ossified		✔
floor of middle ear cavity at a lower level	✔	
nasal cavities at a higher level	✔	
frontal and parietal eminences more prominent	✔	
greatest skull width just above mastoid process of temporal bone		✔

Quote: "Three general processes . . . constitute 'growth': (1) size increase, (2) remodeling, and (3) displacement."

"The reason a bone must undergo such widespread remodeling is because its regional parts become 'relocated' as the whole bone enlarges. The palate, for example, is relocated in a constantly inferior direction. About half of its periosteal surfaces (the nasal side) is resorptive, and about half (oral side) is depository. What at one time was the maxillary arch in the young child becomes directly remodeled into the nasal region at later ages."

D.H. Enlow, 1978 (Whitaker & Randall)

FIGURE 4.23 RADIOGRAPHIC LANDMARKS IN CEPHALOMETRY.

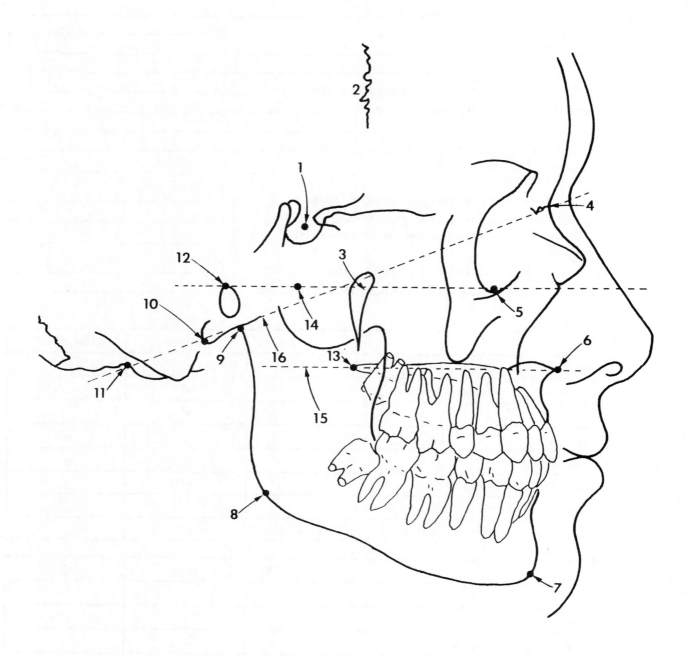

Identify:

1. sella turcica (center)
2. coronal suture
3. pterygomaxillary fissure
4. nasion
5. orbitale
6. anterior nasal spine

7. gnathion
8. gonion
9. articulate
10. basion
11. Bolton point
12. porion

13. posterior nasal spine
14. central ray
15. palatal plane
16. base plane
   (cranial base plane)

Quote: ". . . superimposing serial radiographic tracings on the cranial base does not demonstrate the actual, complex growth changes that occur in each of the many regions of the face as all enlarge in an interrelated, composite manner."

D.   H. Enlow, 1978 (Whitaker and Randall)

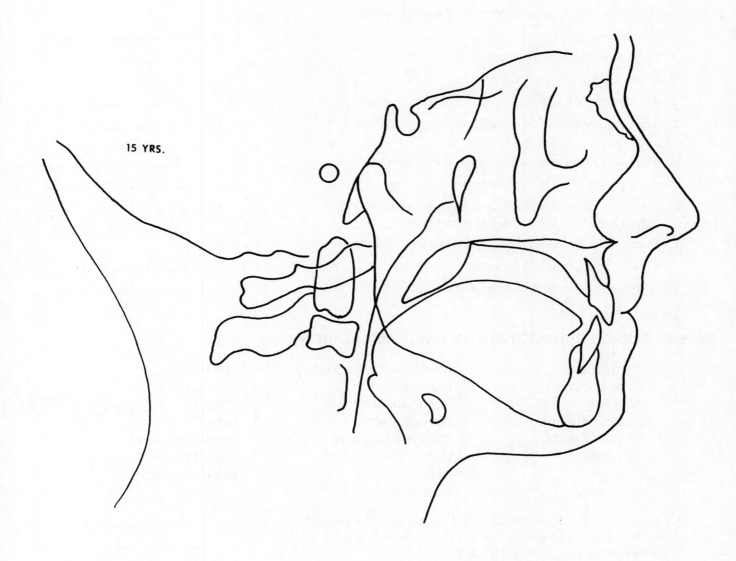

15 YRS.

Because the tracing of the lateral head x-ray shown above has been reduced in size, your measurements will not reflect the actual size but the percentage derived will be unchanged.

Locate and label: 1. anterior nasal spine (ANS)
                       2. posterior nasal spine (PNS)
                       3. palatal plane, *and continue it to the* pharyngeal wall (Ph)

Measure: 1. the length of the soft palate
                    2. the depth of the pharynx at the level of the palatal plane

Determine the adequacy of soft palate tissue for velopharyngeal closure by means of the following equation:

$$\frac{\text{distance between PNS and Ph}}{\text{length of soft palate}} = \underline{\hspace{2cm}} = \underline{\hspace{1cm}} \%$$

See norms in Table 4-4, text page 286.

No. 4-31 RESEARCH TECHNIQUES
Text pages 291–293

1. Name the device most appropriate for determining
   a. intraoral pressure
   b. air flow through the vocal tract
   c. muscle activity
   d. pulmonary function

2. The primary limitation of radiographic research is _____.

3. The x-ray microbeam technique reduces both _____ and _____.

4. Computer generated displays of linguapalatal contacts during speech utilize a technique known as _____.

5. Name the articulation tracking system described below:
   a. is reflected, much like an echo
   b. responds electrically to distortion

1.
a. _____
b. _____
c. _____
d. _____

2. _____

3. _____
_____

4. _____

5.
a. _____
b. _____

No. 4-32 SPEECH PRODUCTION (VOCABULARY REVIEW)

articulation	fundamental frequency	resonance
complex tone	harmonic	spectrum
damping	laryngeal buzz	stop
forced vibration	laryngeal tone	tone
formant bands	natural frequency	unvoiced sound
formants	partial	voiced sound
fricative	period	volume velocity
frequency	overtone	

1. a complex tone whose fundamental frequency is determined by the vibratory rate of the vocal folds, but which has not been modified by the vocal tract (2)

2. any physical component of a complex tone

3. the number of repetitions of a periodic process per unit time, usually Hz (cycles per second)

4. the rate of flow of the medium through a specified area due to a sound wave

5. a physical component of a complex sound having a frequency higher than that of the basic frequency; sometimes used in place of harmonic

6. causing a decrease in amplitude of successive waves or oscillations

7. a speech sound generated when the articulators completely impound the outward flow of air and then suddenly release it

8. a speech sound produced when the vocal folds are vibrating

9. adjustment of the shape and thus the acoustical properties of the vocal tract

1. _____
_____

2. _____

3. _____

4. _____

5. _____

6. _____

7. _____

8. _____

9. _____

10. a partial whose frequency is an integral multiple of the fundamental frequency

    10. _____

11. a frequency of free oscillation of a body or system

    11. _____

12. a structure's absorption and emission of energy at the same frequency band

    12. _____

13. a sound wave produced by the combination of simple sinusoidal components of different frequencies

    13. _____

14. the greatest common divisor of the component frequencies of a period wave or quantity

    14. _____

15. a sound sensation having pitch

    15. _____

16. induced oscillation of a body at an unnatural frequency

    16. _____

17. prominent resonances of the vocal tract

    17. _____

18. a speech sound generated when the air stream passes through a constriction of the vocal tract

    18. _____

19. the distribution in frequency of the magnitudes (and sometimes phases) of the components of the wave

    19. _____

20. the time required for an oscillating body to make one complete oscillating or vibratory cycle

    20. _____

21. a speech sound produced when the vocal folds are not vibrating

    21. _____

22. regions of prominent energy distribution in a speech sound; broad-band resonant frequencies

    22. _____

No. 4-33 SPEECH PRODUCTION; RESONANCE
Text pages 294–295

1. Do you ever hear a glottal or laryngeal tone? Explain.

    1. _____

    _____

2. Short-duration bursts of air which are released into the vocal tract cause vibration of *the air column within the vocal tract / the vocal tract itself / both.*

    2. _____

3. The vibrations generated within the vocal tract

    3.

    a. are of *short / long* duration.

        a. _____

    b. are the product of _____.

        b. _____

    c. are *slightly / highly* damped.

        c. _____

    d. are modified by structures located between the _____ and the _____.

        d. _____

        _____

4. What aspects of voiced speech sounds are determined by vocal fold vibration?

4. _____

_____

_____

5. A tuning fork not only absorbs energy at a specific frequency, it also radiates energy at that frequency, or in other words, _____ to that frequency.

5. _____

6. The resonant frequencies of a vibrating air column can be changed by modifying the cavity's _____ and

_____.

6. _____

_____

Characteristics of a System Vibrating at:	Natural Frequency	Unnatural Frequency
damping forces are slight	✔	
damping forces are great		✔
when outside force is removed vibrations will cease abruptly		✔
when outside force is removed vibrations will continue	✔	

Conditions	Rate of Vocal Fold Vib.	Acoustic Prop. of Vocal Tract
you sing "ah" as you go up the musical scale	(Changed) Unchanged	Changed (Unchanged)
you sing a different vowel on each note as you go down the musical scale	(Changed) Unchanged	(Changed) Unchanged
you sing a variety of vowels at 256 Hz	Changed (Unchanged)	(Changed) Unchanged
you maintain an "ah" sound at 256 Hz for ten seconds	Changed (Unchanged)	Changed (Unchanged)

Notes: *Vocal quality,* as described in Chapter 3, is determined at the laryngeal level.

Vowel quality, which refers to phonemic differences such as [i] [o] [a] [e], is determined by the frequency or frequencies at which the air column resonates.

Voice quality, which may refer to the distinctive characteristics of a particular voice, is probably influenced by the uniqueness of an individual's vocal tract and by subtle, nonphonemic variations in the articulation of complex sounds.

The terms laryngeal tone, laryngeal buzz, glottal tone, vocal fold tone, sound source, and voice source are used interchangeably.

Quote: "The source of voice is vibration of the vocal folds. The vocal-fold tone is the type of complex sound that may be viewed as the summation of a harmonic series of simple tones. The spectrum of the vocal-fold tone is raw material, not yet differentiated phonetically. It is usually agreed that this spectrum, if observable, would show most of the energy in the fundamental, and progressive reduction from the harmonic upward. By the time we see (hear) the spectrum, it has been modified selectively by the cavities and conduits of the articulatory mechanism."

Grant Fairbanks, 1960

No. 4-34 SOURCE-FILTER THEORY OF SPEECH PRODUCTION
Text pages 295–299

The *sound pressure spectrum* P(f) at some distance from the lips is the product of:

U(f) is the amplitude vs. frequency characteristics of the source.

H(f) is the frequency-selective gain function of vocal transmission.

R(f) is the radiation characteristics at the lips.

| | denotes concern with only magnitude of function.

(f) denotes function of frequency.

$$|P(f)| = |U(f)| \cdot |H(f)| \cdot |R(f)|$$

Relate the above equation to Fig. 4-116, text page 296.

Aspects of Speech Production Related to:	U(f)	H(f)	R(f)
transfer function of the vocal tract		✔	
radiation of air pressure wave at the lips			✔
characteristics of the sound source	✔		
represented by resonance curves		✔	
most directly related to perceived differences among speech sounds		✔	
source spectrum	✔		
fundamental frequency and harmonics	✔		
glottal or laryngeal tone	✔		
vocal tract configuration		✔	
power supply and oscillator	✔		
output			✔
volume velocity spectrum generated by source	✔		
formant frequencies		✔	
largely determines vowel quality		✔	
largely determines vocal quality	✔		
resonances		✔	
sound pressure pattern			✔

1. As fundamental frequency increases, the number of harmonics or overtones *increases / decreases.*

2. The voice-source spectrum demonstrates that the relationship between frequency and amplitude of the harmonics is
   direct / inverse; in other words, amplitude decreases uni-
   formly as frequency *increases / decreases.*

1. _____

2. _____
   _____

3. See Fig. 4-119, text page 297. FORMANTS: First   Second   Third

   a. If the vocal tract were half as long as it is
      in the figure, the frequencies of the formants
      would be: 3. a. _____

   b. If the vocal tract were twice the length shown,
      the frequencies of the formants would be: b. _____

Fundamental Wave Equation

$$f = \frac{V}{\lambda} \text{ frequency of Formant}_1 = \frac{\text{velocity of sound}}{\text{wavelength of } F_1} = \frac{340 \text{ meters } (34{,}000\,cm)/\text{second}}{4 \times \text{length of tube}}$$

$$\text{frequency of Formant}_2 = 3 \times \text{frequency of } F_1 \qquad \text{frequency of Formant}_3 = 5 \times \text{frequency of } F_1$$

4.	Length of tube closed at one end	Wavelength of F_1	Formant Frequencies F_1	F_2	F_3
a.	20 centimeters	cm			
b.	10 centimeters	cm			
c.	40 centimeters	cm			

5. The attenuation of a partial of the sound source is directly
related to the proximity of the frequencies of the partial
and the _____. If the two frequencies are the same,
amplitude at the lips will be *maximal / minimal*.

5. _____
_____

6. Why do men tend to have lower formant
frequencies than women?

6. _____
_____

7. Air molecule displacement at the lips is greater for *high /
low* intensity sounds and for *high / low* frequency sounds.

7. _____
_____

Components of a Complex Tone	125Hz	250Hz	375Hz	500Hz
partials	✔	✔	✔	✔
overtones		✔	✔	✔
harmonics	✔	✔	✔	✔
fundamental frequency	✔			
first harmonic	✔			
second overtone			✔	
third harmonic			✔	

Question: How would a voice sound if the vocal tract length were that of an adult male
and the vocal fold vibratory rate were that of a child?

No. 4-35 VOWEL CLASSIFICATION
Text pages 300–301

1. Why are the cardinal vowels useful? 1. _____

_____

2. On the cardinal vowel diagram shown below (in the parentheses next to the phonetic symbol) write the word containing that vowel. *hoed, hard* (New England dialect, [r] not pronounced), *hod, heed, who'd, head, hawed, hayed*

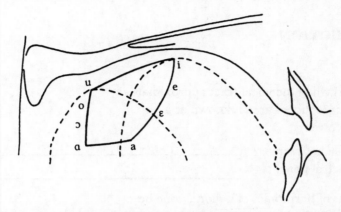

The word *cardinal* is derived from compass terminology in which north, south, east, and west are the cardinal points and northwest, northeast, southwest, and southeast are the collateral points. Although we commonly plot eight vowels on the cardinal vowel diagram, some phoneticians consider the four vowels which are plotted in the corners of the diagram the true cardinal vowels, and refer to the others as collateral vowels. *Underline the true cardinal vowels.*

3. As you say the following vowels try to determine what your articulators are doing and fill in the chart accordingly. No *right answers* required.

Vowel	Major Constriction Front Central Back	Tongue Height High Mid Low (Close) (Open)	Tension Tense Lax	Lip Shaping Round Neutral Spread
[i] beat				
[I] bit				
[e] bait				
[ɛ] bet				
[æ] bat				
[u] fool				
[ʊ] full				
[ɔ] fall				
[o] hoed				
[ʌ] hum				
[ɑ] calm				
[ə] ado				

Try producing "experimental" vowels with varying combinations of tongue height, position, tension, and lip rounding.

No. 4-35 cont'd

4. Longer duration and greater acoustic intensity are generally characteristic of *tense / lax* vowels.

4. _____

5. The smooth transition of one vowel sound to another within
one syllable occurs in a/an _____.

5. _____

No. 4-36 VOWEL PRODUCTION
Text pages 301–303

1. The vowel produced when the vocal tract approximates a
tube of uniform cross-sectional area is known as a/an
_____ vowel.

1. _____

2. What is the difference between
a formant and a spectral peak?

2. _____

_____

3. Can the distributions of harmonics of a glottal tone be
modified without changing the rate of vocal fold vibration?

3. _____

4. Do the spectral peaks of a vowel change if the vowel is
produced at a number of different fundamental frequencies?

4. _____

5. If different vowels are produced at the same fundamental
frequency is there a change in the distribution of formants of the vocal tract?

5. _____

6. What are the three parameters of the vocal tract that
can be controlled by the articulators?

6. _____

_____

_____

7. Are the frequencies of the formants directly or inversely
proportional to the length of the vocal tract?

7. _____

8. How may the vocal tract be lengthened?

8. _____

_____

9. For vowel production, the primary constrictors of
vocal tract are the _____ and _____.

9. _____

_____

Questions: When you were analyzing your tongue positions for different vowels what did you discover about
this sensory feedback mechanism?

Why would it be difficult to learn to produce vowels you could not hear?

Say the word list (no. 1) once very slowly and then very rapidly. How does rate affect articulatory
movements?

No. 4-37 PLOTTING THE FREQUENCIES OF VOWEL FORMANTS
Text pages 302–303

Excerpts of spectrographic analyses of the vowels in the words shown beneath the samples.

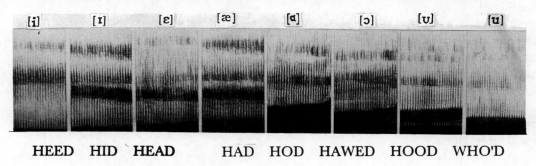

	[i]	[ɪ]	[ɛ]	[æ]	[ɑ]	[ɔ]	[ʊ]	[u]
	HEED	HID	HEAD	HAD	HOD	HAWED	HOOD	WHO'D

Formant Frequencies (Hz)

F_1	270	390	525	650	730	570	440	300
F_2	2300	2000	1850	1725	1100	840	1025	870

Plot the frequency of formant (2) against the frequency of formant (1) for each vowel, with its phonetic symbol. When the symbols are connected by a continuous line the resulting 'vowel loop' should resemble the shape of the vowel quadrilateral shown in Figure 4-125, text page 301. Compare your vowel loop with the one shown in Figure 4-128, text page 303. Because the centers of each gray bar on the right of the spectrograms are separated by 500 Hz, the values for formant locations are only approximate.

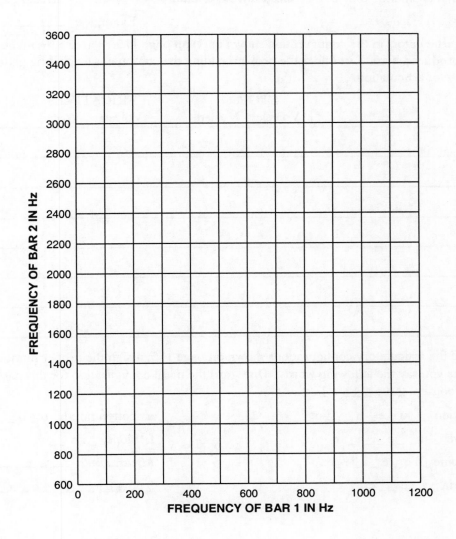

Key:
1. lips (labial)
2. teeth (dental)
3. alveolar ridge (alveolar)
4. hard palate (prepalatal)
5. hard palate (palatal)
6. soft palate (velar)
7. uvula (uvular)
8. pharynx (pharyngeal)
9. vocal folds (glottal)

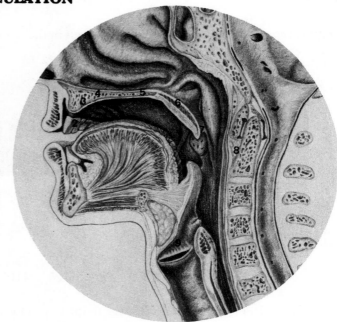

1. Of the nine places of articulation shown above, which two are not normally used in English?

  1. _____
  _____

2. Two consonants differing only in the voiced-unvoiced feature are called

  2. _____
  Examples: _____

3. Place your articulators in the configuration shown in (1) on page 175. Three consonants including [p] as in *pie* can be produced here. List under the proper heading the consonants (excluding glides) that can be made at each place of articulation.

Place of Articulation	STOPS Voiceless	Voiced	FRICATIVES Voiceless	Voiced	NASALS
Labial					
Labiodental					
Dental					
Alveolar					
Palatal					
Velar					
Glottal					

4. Referring to the articulatory configurations shown on page 175, assign the proper number to each consonant as you say the following words. Disregard the nasal configuration for this particular exercise.
Example: footstep, 2 o o 4 4 4 e 1

  a. stupefaction _ _ u _ e _ a _ _ i o _

  b. subjective _ u _ _ e _ _ i _ e

  c. biomedicine _ i o _ e _ i _ i _e

  d. kinesthesia _ i _ e _ _ e _ i a

  e. cottonmouth _o _ o _ _ o u _

  f. biopsy _ i o _ _ y

  Mystery words: _____

  3 i 6 4 6 i 4 e 4 / 4 o 1 a 8 a w 6

FIGURE 4.24 ARTICULATORY CONFIGURATIONS FOR THE PRODUCTION OF CONSONANTS.

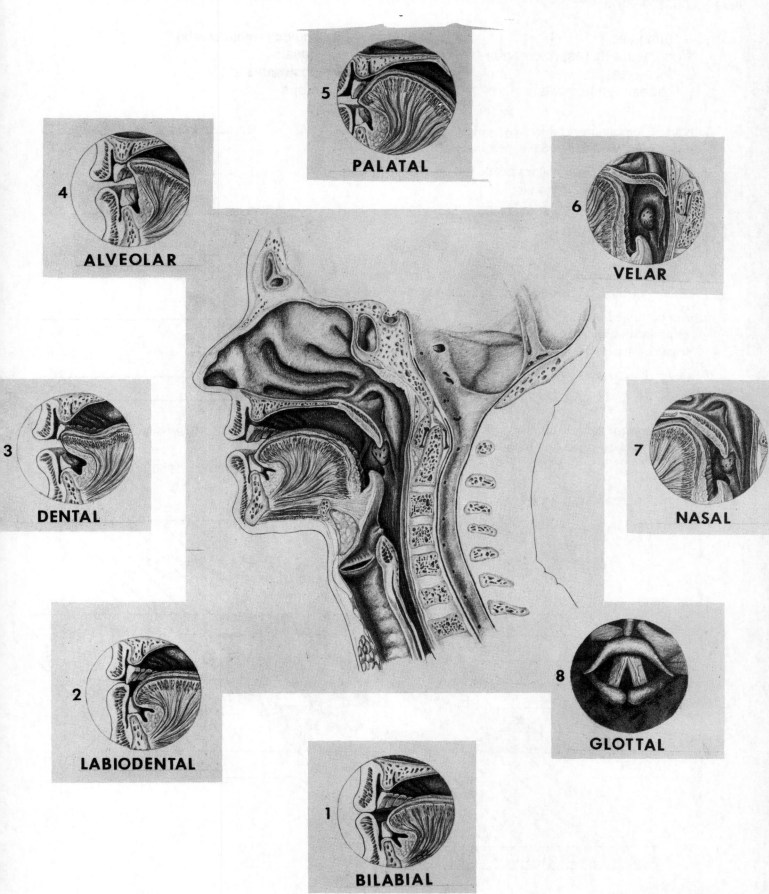

5 PALATAL

4 ALVEOLAR

6 VELAR

3 DENTAL

7 NASAL

2 LABIODENTAL

8 GLOTTAL

1 BILABIAL

affricates
continuants (approximants) (sonorants)
fricatives
glides (semi-vowels)

liquids (semi-vowels)
nasals
obstruents
stops

1. voiced consonants produced by complete constriction of the oral cavity and with an open nasopharyngeal port

1. _____

2. complete closure of vocal tract followed by a fricative-like release of impounded air

2. _____

3. complete closure of vocal tract followed by a sudden release of impounded air

3. _____

4. [l] [r]

4. _____

5. noise resulting from the increased velocity of the air stream as it flows through an incomplete constriction of the vocal tract

5. _____

6. similar to and usually preceding vowels, but production requires movement of an articulator and greater vocal tract constriction; [w] as in wake, [j] as in you

6. _____

7. glides, liquids, and nasals; flow of breath is not obstructed

7. _____

8. stops, fricatives, affricates; flow of breath is obstructed

8. _____

Quote: "In *stops* the oral channel is first closed (stop phase) and then opened abruptly (plosive phase) to release a puff (aspiration), but plosive phase and aspiration phase are variable."

Grant Fairbanks, 1960

FIGURE 4.25 CONSONANT CHART. (variation on a theme by Daniloff et al.)

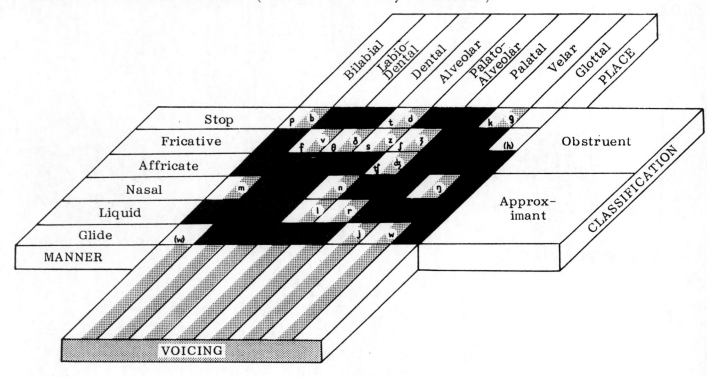

Description	Vowels	Consonants
62 percent of English speech sounds		✔
38 percent of English speech sounds	✔	
less constrictive	✔	
convey more "information"		✔
usually longer in duration	✔	
more often initiate and terminate syllables		✔
more responsible for the transitory nature of speech		✔

Description — Consonants:	Voiced	Unvoiced
higher intraoral pressure		✔
longer voice-onset-time		✔
formant transition occurs before onset of voice		✔
formant transition occurs after onset of voice	✔	
aspirated		✔

1. Voice-onset-time is the time interval between the instant of _____ and the instant _____ begins.

   1. _____

      _____

2. Say the word *top*, first with an unaspirated [p] and then with an aspirated [p]. Describe the differences between the two.

   2. _____

      _____

3. Why is the first formant lowered in nasals [m] and [n], but not in [ŋ] as in sing?

   3. _____

      _____

4. Consonants are specified by *resonances/ antiresonances / both*. Vowels are specified by _____.

   4. _____

      _____

Quote: "Theoretically, the addition of a side-branch resonator such as the nasal cavity will introduce *antiresonances* in the transfer function of the complete resonator system. Asymmetries in the nasal channel will also introduce antiresonances. Antiresonances have the opposite effect of resonances, in that they produce spectral minima or valleys as opposed to spectral maxima or peaks. Similar to resonances, however, they are characterized by specific frequency locations and bandwidths. They may have the effect of partially canceling a resonant mode in a nearby frequency location."

J. Shoup, N. Lass, D. Kuehn, 1982 (Lass et al.)

1. The ultimate criteria for meeting the target of correct sound production is *auditory / articulatory.*

    1. _____

2. Semantic differences are conveyed by *allophones / phonemes.*

    2. _____

3. The features of a sound segment that were in preceding segments and/or continue into the following segments help account for the *distinct / indistinct* boundaries between phonemes.

    3. _____

4. Prosadic features which impart stress, intonation and inflection to speech are the _____ elements of speech.

    4. _____

5. The consonant to vowel sequences, the vowel to consonant sequences, and diphthongs are characterized by formant _____.

    5. _____

6. Each plosive consonant has its characteristic _____ frequency.

    6. _____

7. The overlapping of articulatory gestures is called _____.

    7. _____

8. The spreading of features often occurs with nasalization.

    8. T   F

9. The role of auditory feedback is more important in *children / adults.* Why? _____

    9. _____

    _____

10. In the process of motor feedback, the difference between the afferent and efferent neural impulses is weighed by a _____. Transmission of this information back to the lower motor neuron facilitates_____ movement.

    10. _____

    _____

Question: How does your understanding of coarticulation affect your methods of remediating phonological problems or misarticulations?

Quote: "Making a sound is not making speech; no one can talk through an oboe. It is necessary to articulate, to join together, before the basic note or notes are transformed into talk."

Anthony Smith, 1986

Chapter 5
The Nervous System

No. 5-1 INTRODUCTION AND ORGANIZATION (VOCABULARY)
Text pages 319–321

afferent
axon
brain
central nervous system
dendrite

efferent
endocrine system
nerve
nerve tract
nervous system

neuron
neurotransmitter
peripheral nervous system
synaptic cleft

1. system mediating primarily internal body processes; usually slow to react

1. _____

2. system mediating primarily observable behavior; usually quick to react

2. _____

3. located within the bony confines of the skull and spinal column

3. _____

4. located outside the bony confines of the skull and spinal column

4. _____

5. "center of the nervous system"

5. _____

6. a specialized conducting cell of the nervous system

6. _____

7. chemical agent inhibiting or facilitating neural transmission

7. _____

8. a nerve process; usually carries impulses away from the cell body

8. _____

9. a nerve process; carries impulses toward the cell body

9. _____

10. carrying toward

10. _____

11. carrying away

11. _____

12. composed of a bundle of axons which may have a variety of functions

12. _____

13. composed of axons having one specific function

13. _____

14. the actual space between two neurons in a chain

14. _____

No. 5-2 THE BRAIN AND THE BRAIN STEM

BRAIN

FOREBRAIN
    cerebral hemispheres
    basal ganglia
        caudate nucleus
        lenticular nucleus

HINDBRAIN
    cerebellum

BRAIN STEM

FOREBRAIN
    thalamus
    hypothalamus

MIDBRAIN
    cerebral peduncles
    corpora quadrigemina

HINDBRAIN
    medulla oblongata
    pons

Note: The cerebral hemispheres are always considered part of the brain and the medulla oblongata and the pons are always considered part of the brain stem, but the assignment of the basal ganglia, the diencephalon, and the midbrain to either the brain or the brain stem is subject to interpretation.

Embryological development of the nervous system is summarized on page 282. Adult derivatives of primary brain vesicles are shown on pages 283 and 285.

No. 5-3 INTRODUCTION TO THE CENTRAL NERVOUS SYSTEM (VOCABULARY)
Text pages 321–325

arachnoid mater	cerebral aqueduct	hypothalamus	pons
basal ganglia	cerebrospinal fluid	lobes of the brain	spinal cord
brain stem	cortex	medulla oblongata	sulci
cerebellum	dura matter	meninges	thalamus
cerebrum	gyri	pia mater	

1. the enlarged region where the spinal cord merges into brain

    1. _____

2. the part of the hindbrain occupying the fossa behind the brain stem; integrating center for coordination and equilibrium

    2. _____

3. an upward continuation of the medulla oblongata which functions as a bridge between the two hemispheres of the cerebellum

    3. _____

4. gray matter covering the cerebellum and cerebrum

    4. _____

5. major relay and integration center for sensory information going to the cerebral cortex

    5. _____

6. fluid-filled cavity in the midbrain; connects fourth ventricle of hindbrain to ventricles in forebrain

    6. _____

7. made up of nuclei controlling visceral and metabolic functions, sleep, temperature, water balance, etc.

    7. _____

8. continuation of the lower end of the medulla oblongata

    8. _____

9. fluid from fourth ventricle; occupies space between the arachnoid and pia mater and circulates around brain and spinal cord

    9. _____

No. 5-3 cont'd

10. the paired convoluted cerebral hemispheres united by the corpus callosum; largest portion of the human brain

10. _____

11. convolutions on the surface of the cerebral hemispheres

11. _____

12. grooves or furrows; depressions separating the convolutions of the brain

12. _____

13. largely demarcated by the lateral and central sulci; names derived from overlying cranial bones

13. _____

14. an internal layer of multilayered gray and white matter within the cerebral hemispheres

14. _____

15. comprising the diencephalon, mesencephalon, pons, and medulla oblongata; a major integrating center of sensory, motor, and bodily functions

15. _____

16. three layers of membranous connective tissue enveloping the brain and spinal cord

16. _____

17. the outermost, toughest, most fibrous meninx; hard + mother

17. _____

18. the middle meninx, a weblike membrane loosely investing the brain; spider's web + mother

18. _____

19. the innermost meninx, a vascular connective membrane closely investing the brain; gentle, tender + mother

19. _____

Note: The way in which the *afferent* components of nervous system are *affected* by the body's internal and external environment determines what action will be *effected* by the *efferent* components.

Description Matter:	Gray	White
generally unmyelinated	✔	
generally myelinated		✔
outer layer of cerebellum	✔	
outer layer of spinal cord		✔
outer layer of cerebral hemispheres	✔	
inner layer of spinal cord	✔	
inner layer of cerebrum		✔
inner layer of cerebellum		✔

Quote: "The brain at birth is prepared to take on a number of its functions, such as regulating heartbeat and breathing and body temperature. But development of a mind requires interaction of the organism with its physical, intellectual, emotional and social environment. The human brain is well prepared to begin taking on this function. Experience molds it, strengthening some connections and some circuits. The brain is prepared to sense sights, sounds, textures, and smells, and it is prepared to remember them. As the brain develops it is prepared to compare new sights and sounds with the ones in memory and to begin groupings of them on the basis of shared characteristics. Thus the brain learns and shapes a mind."

F. Bloom and A. Lazerson, 1988

No. 5-4 THE PERIPHERAL NERVOUS SYSTEM: INTRODUCTION
Text pages 325–326

1. Where is the peripheral nervous system located?

   1. _____

2. What are the two major divisions of the peripheral nervous system?

   2. _____
   _____

3. On a functional basis, what are the three types of cranial nerves?

   3. _____
   _____
   _____

4. All spinal nerves are of what type?

   4. _____

5. The nuclei of origin of the motor cranial nerves are located in the _____.

   5. _____

6. Sensory cranial nerves arise from groups of cells *within / outside* the brain.

   6. _____

7. The motor fibers of the spinal nerves arise in the cell bodies in the ventral column of *gray / white* matter in the _____.

   7. _____
   _____

8. The sensory fibers of the spinal nerves arise in cell bodies in ganglia *inside / outside* the spinal cord.

   8. _____

9. The groups of nerve cells in the brain upon which the central processes of the sensory cranial nerves arborize are called nuclei of _____.

   9. _____

10. The name of a cranial nerve may indicate its _____, _____, or _____.

    10. _____
    _____
    _____

11. Which portion of the peripheral nervous system controls the internal environment of the body?

    11. _____

12. Which portions of the peripheral nervous system react and adjust to the external environment?

    12. _____

No. 5-5 THE AUTONOMIC NERVOUS SYSTEM: INTRODUCTION
Text pages 326–327

1. What is another name for the autonomic nervous system?

   1. _____

Description Fibers:	Preganglionic	Postganglionic
axons of ganglion cells		✔
axons of nerve cells in the brain stem and spinal cord	✔	
follow pathways of certain cranial nerves and ventral roots of spinal nerves	✔	
continue to ganglia outside the central nervous system	✔	
supply the viscera		✔
supply smooth and cardiac muscle		✔

No. 5-5 cont'd

Description Divisions of ANS:	Sympathetic	Parasympathetic
craniosacral		✔
thoracolumbar	✔	
ganglia located at or near organ being supplied		✔
long nerve trunk with ganglia at regular intervals flanking either side of the vertebral column	✔	
branches of the nerve trunks form plexuses	✔	
white rami communicantes connect nerve trunks with ventral rami of spinal nerves	✔	
supplies viscera of the head and neck	✔	✔
supplies viscera of the pelvis	✔	✔

No. 5-6 FUNCTIONAL ANATOMY OF THE CENTRAL NERVOUS SYSTEM: THE MENINGES
Text pages 327–329

1. The three layers of non-nervous tissue surrounding the brain and spinal cord are called _____.
   The spaces between these layers are filled with _____ fluid.
   What are the functions of this fluid? _____

   1. _____
      _____

2. The two layers of dura mater separate to enclose the _____ sinuses which drain _____ and _____ from the brain.

   2. _____
      _____
      _____

3. Formations created by the close contact of the arachnoid mater and pia mater at the height of the convolutions of the brain are called _____.

   3. _____

4. Extensive spaces between the arachnoid and brain are called subarachnoid _____.

   4. _____

5. Small tufts, acting as one-way valves permitting cerebrospinal fluid to be resorbed into the venous bloodstream, are called arachnoid _____,

   5. _____

6. Strands of connective tissue resembling little beams or cross bars are called _____. They arise from the pia mater, attaching it to the _____ mater.

   6. _____
      _____

7. Why is the pia mater pink?   7. _____

8. The major sites of the formation of cerebrospinal fluid are the _____ plexuses located in the _____.

   8. _____
      _____

Description Folds of Meningeal Dura:	Falx Cerebri	Tentorium Cerebelli	Falx Cerebelli
separates cerebellar hemispheres			✔
separates cerebral hemispheres	✔		
separates cerebellum from occipital lobes		✔	
in the longitudinal fissure	✔		
in the transverse fissure		✔	

Questions: What is meningitis?

What keeps the brain from "bouncing around" inside the skull?

If the middle meningeal artery is damaged, blood will flow between the outer surface of the dura and the inner surface of the skull. Speculate on the consequences.

FIGURE 5.1 SCHEMATIC LATERAL VIEW OF THE BRAIN, INCLUDING THE CEREBRUM, CEREBELLUM, PONS, MEDULLA OBLONGATA, AND SPINAL CORD.

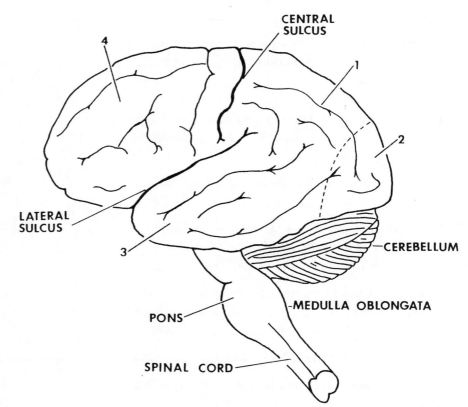

On the basis of the relationship of the brain to the bones of the skull

Identify:

1. parietal lobe
2. occipital lobe
3. temporal lobe
4. frontal lobe

1. The motor area for speech is in the frontal lobe at the junction of the _____ and _____ fissures.

   1. *lateral*
      central

2. The lobe of the brain which lies at the bottom of the lateral fissure is the _____ lobe.

   2. *insular*

3.
   corpus callosum   postcentral gyrus (parietal)
   left angular gyrus (parietal)   precentral gyrus (frontal)
   left inferior frontal gyrus (frontal) temporal operculum(temporal)
   limbic system(not a true lobe)

   a. may be associated with short-term memory

   b. Broca's speech area

   c. provides communication between hemispheres

   d. primitive structure associated with "animal behavior"

   e. associated with reading and writing

   f. associated with common motor pathway to skeletal muscles

   g. cortical center for hearing

   h. damage may cause alexia ("word blindness")

   i. damage may cause aphasia (language problems)

   j. damage may cause agraphia (writing problems)

   k. consists of medial margins of temporal, frontal and parietal lobes

   3. a. *limbic system*
      b. *left inf. frontal gyrus*
      c. *corpus callosum*
      d. *limbic system*
      e. *left angular gyrus*
      f. *precentral gyrus*
      g. *temporal operculum*
      h. *left angular gyrus*
      i. *left inf. frontal gyrus*
      j. *left angular gyrus*
      k. *limbic system*

Description/Structures Interior fibers:	Projection	Commissural	Association
connect cortical regions of two hemispheres		✔	
convey impulses between remote regions and cerebral cortex	✔		
interconnect cortical regions in same hemisphere			✔
in white matter of cerebrum	✔	✔	✔
uncinate and arcuate fasciculi; cingulum			✔
corpus callosum, anterior commissure		✔	
form corona radiata and internal capsule	✔		
long and short bundles of fibers			✔
contained in semioval center	✔	✔	✔

Quote: "Each of the two million nerve fibers in the corpus callosum fires a nerve impulse on the average of 20 times a second, which means that each second—right now, in fact—there are four billion nerve impulses firing back and forth between your left and right hemispheres."

N. McAleer, 1985

FIGURE 5.2 DIAGRAMMATIC CEREBRAL FUNCTIONS. (Some are well established, others speculative or hypothetical.)

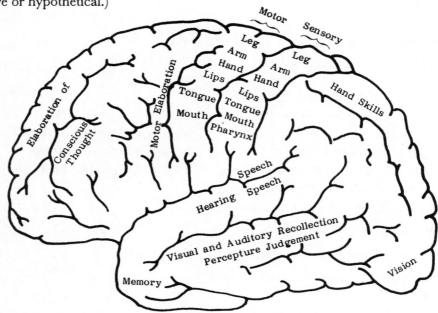

FIGURE 5.3 SCHEMATIC SAGITTAL SECTION THROUGH THE BRAIN, SHOWING GENERAL PLAN OF THE VENTRICLES.

Identify:

1. lateral ventricle
2. third ventricle
3. cerebral aqueduct
4. fourth ventricle
5. central canal

The shaded area above the lateral ventricle is the corpus callossum.

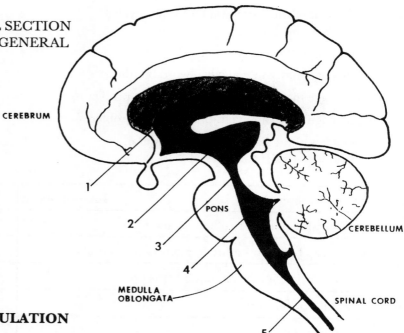

No. 5-8 CEREBROSPINAL FLUID CIRCULATION
Text pages 336–337

1. Summary of circulation. Cerebrospinal fluid

    a. is produced in the ventricles by the specialized epithelial cells constituting the _____ plexus.

    b. begins its circulation in the _____.

    c. then diffuses into the _____ spaces.

    d. enters the cisterns and flows around the _____.

    e. is allowed to pass into the venous bloodstream by the _____ granulations which project into the _____ sinus.

1.

   a. _____

   b. _____

   c. _____

   d. _____

   e. _____

   _____

2. Elevated cerebrospinal fluid pressures may result if the
    fluid cannot be _____ or _____.

    2. _____

    _____

3. What are some of the symptoms
    of elevated cerebrospinal fluid
    pressure?

    3. _____

    _____

FIGURE 5.4 THE VENTRICULAR SYSTEM OF THE BRAIN.

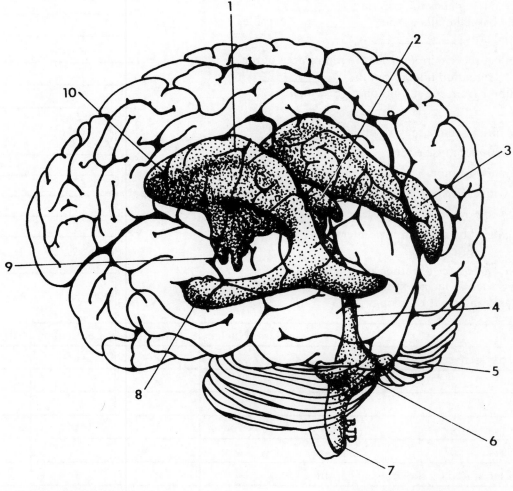

Identify:
1. body of lateral ventricle
2. third ventricle
3. posterior horn
4. cerebral aqueduct
5. lateral recess (4th vent.)

6. fourth ventricle
7. central canal (spinal cord)
8. inferior horn
9. optic recess
10. anterior horn

Question: How does the cerebrospinal fluid provide a means to reduce the effective weight
of the brain?

Note: For diagnostic purposes cerebrospinal fluid is drawn from a lumbar puncture, usually between L3 and L4.
Measurement of spinal fluid pressure may reveal obstruction of circulation due to hydrocephalus, injury,
brain tumor, or other abnormalities. Analysis of protein, gamma globulin, glucose, and chloride levels,
and white and red blood cell counts may aid in the diagnosis of tumors, cranial hemorrhages, multiple
sclerosis, viral or bacterial meningitis, and other central nervous system infections.

1. The basal ganglia nuclei are composed of _____ matter.

    1. _____

2. The thin layer of gray matter located between the lateral margin of the putmen and the cortex of the insula is called the _____, meaning an enclosed space or barrier.

    2. _____

   Word Association: Fear of enclosed spaces is what kind of phobia?

3. The projection fibers from the cerebral cortex converge toward the basal ganglia as the corona _____. They enter the basal ganglia as the _____ capsule, and leave as the crus _____.

    3. _____
       _____
       _____

4. The basal ganglia receive input from the motor area of the _____, radiation fibers from the _____, and afferent fibers from the substantia nigra which synthesizes _____:

    4. _____
       _____
       _____

5. Lesions of the basal ganglia may result in

    5.

   a. *increased / decreased* muscle tone.

    a. _____

   b. *intention / resting* tremor.

    b. _____

   c. problems with *voluntary / involuntary* movement.

    c. _____

   d. *athetosis / spasticity.*

    d. _____

   e. a disease caused by a dopamine deficiency, _____ disease.

    e. _____

   f. *exaggeration / diminution* of rather automatic movements such as facial expression, or arm swinging associated with walking.

    f. _____

Description	Basal Nuclei:	Caudate	Lenticular	Amygdaloid
with internal capsule form striate bodies		✔	✔	
"having a tail"		✔		
"almond shaped"				✔
"lens-shaped"			✔	
part of the limbic system				✔
except anteriorly, separated by the internal capsule		✔	✔	
continuous with the tail of the caudate nucleus				✔
composed of putamen (L. shell) and globus pallidus (L. pale sphere)			✔	
role in food and water intake; general orientation				✔

Quote: "One aspect of skeletal muscle activity cannot be consciously controlled. Even when a muscle is voluntarily relaxed, some of its fibers are contracting—first one group and then another. Their contraction is not visible, but as a result of it the muscle remains firm, healthy, and constantly ready for action. This state of continuous partial contractions is called *muscle tone*. Muscle tone is the result of different motor units, which are scattered through the muscle, being stimulated by the nervous system in a systematic way.

    If the nerve supply to a muscle is destroyed (as in an accident), the muscle is no longer stimulated in this manner, and it loses tone and becomes paralyzed. It then becomes *flaccid* (flak´sid), or soft and flabby, and begins to atrophy (waste away)."

Elaine N. Marieb, 1994

FIGURE 5.6 SCHEMATIC FRONTAL SECTION OF CEREBRUM.

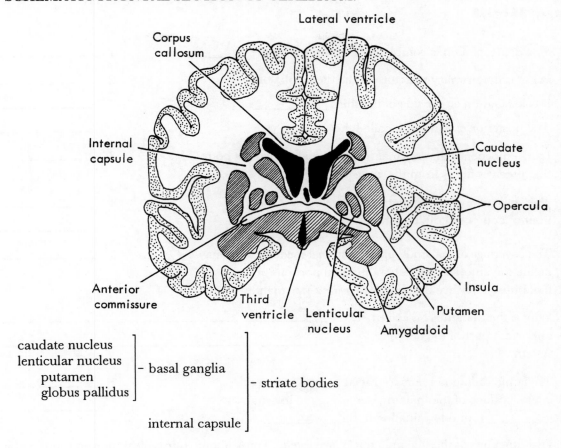

Lateral ventricle

Corpus callosum

Internal capsule

Caudate nucleus

Opercula

Anterior commissure

Insula

Third ventricle

Lenticular nucleus

Putamen

Amygdaloid

caudate nucleus
lenticular nucleus
putamen
globus pallidus
]
- basal ganglia
]
- striate bodies

internal capsule
]

FIGURE 5.6 SCHEMATIC TRANSVERSE SECTION OF CEREBRUM.

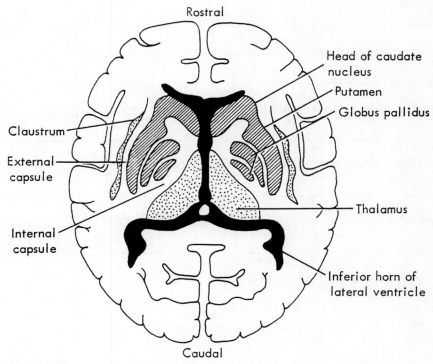

Rostral

Head of caudate nucleus

Putamen

Globus pallidus

Claustrum

External capsule

Thalamus

Internal capsule

Inferior horn of lateral ventricle

Caudal

Ventricles are shown in black.

1. The diencephalon is completely surrounded by the _____.

          1. _____

2. What is the primary structure of the diencephalon?

          2. _____

   It is located on either side of the third _____.

          _____

   The posterior expansion is called the _____.

          _____

3. Located beneath the pulvinar is the medial geniculate body that receives fibers from the *auditory / optic* pathway.

          3. _____

4. The substance of the thalamus is largely *gray / white* matter, composed of 26 pairs of thalamic _____.

          4. _____

          _____

5. The coverings of thalamus and the internal medullary lamina define the anterior, medial, and lateral portions of the thalamus. They are thin layers of *white / gray* matter.

          5. _____

6. Why is the thalamus an extremely important part of the nervous system?

   6. _____

      _____

7. Thalamic radiation refers to tracts emerging from the lateral surface of the thalamus, entering the internal _____, and terminating in the _____.

          7. _____

          _____

8. Five functions of the thalamic nuclei are listed. Think about your typical morning routine for a school day and jot down any sensations, reactions, or experiences that might be examples of those functions.

   a. replaying/processing all sensory input (except olfaction) to cerebral cortex for conscious awareness

_____

   b. perception of crude aspects of pain/temperature/touch (not accurate localization)

_____

   c. imparting pleasantness/noxiousness to sensations (thus influencing emotional response)

_____

   d. maintaining cortical activity for arousal/attention/sleep-wake cycles

_____

   e. relay / integrate input from cerebellum and globus pallidus to motor center

_____

FIGURE 5.7 THE CAUDATE NUCLEUS, THALAMUS, AND AMYGDALOID NUCLEUS SHOWN IN RELATION TO THE CEREBRUM.

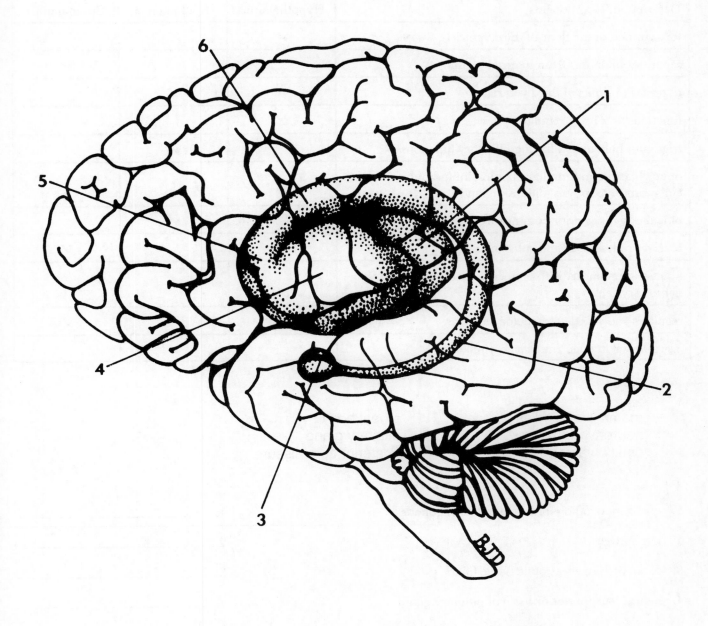

Identify: 1. thalamus
            2. tail of caudate nucleus
            3. amygdaloid nucleus
            4. lenticular nucleus
            5. head of caudate nucleus
            6. caudate nucleus

Note: Having difficulty performing voluntary movements is called *dyskinesia*.

Description/Components	Hypothalamus	Epithalamus	Subthalamus
forms much of the floor of third ventricle	✔		
separates thalamus from internal capsule			✔
at posterior limits of third ventricle		✔	
functions as a body regulator	✔		
trigonum habenulae, having olfactory functions		✔	
optic chiasma, mammillary bodies, infundibulum, tuber cinereum, hypophysis (pituitary gland)	✔		
pineal body, posterior commissure		✔	
receives red nucleus and substantia nigra			✔
functions in concert with endocrine and autonomic nervous systems	✔		
regulates and coordinates motor functions			✔
instrumental in optic and olfactory reflexes		✔	

mammillary bodies   pineal body
neurohypophysis   pituitary gland
optic chiasma   posterior commissure

1. gland functioning in development of gonads   1. _____

2. hypophysis   2. _____

3. X-shaped; receives optic nerve fibers   3. _____

4. neural part (posterior lobe) of pituitary gland   4. _____

5. connect superior colliculi   5. _____

6. part of limbic system; breastlike   6. _____

7. landmark (reference point) in x-ray
    examination of the brain   7. _____

FIGURE 5.8 SAGITTAL SECTION OF THE BRAIN STEM AND CEREBELLUM (top) AND BRAIN STEM AS SEEN FROM BENEATH (bottom) SHOWING STRUCTURES OF THE HYPOTHALAMUS AND EPITHALAMUS.

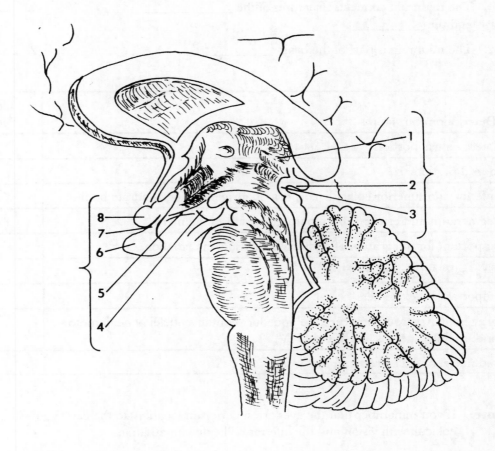

Identify:

1. habenular nuclei

2. pineal body

3. posterior commissure

4. mammillary body

5. tuber cinereum

6. hypophysis (pituitary)

7. infundibulum

8. optic chiasm

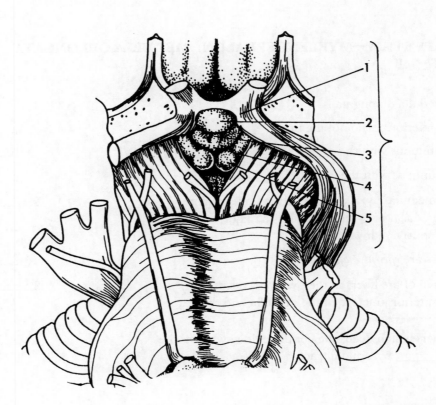

Identify:

1. optic chiasm

2. hypophysis (pituitary)

3. infundibulum

4. tuber cinereum

5. mammillary body

Label:

hypothalamus

epithalamus

No. 5-12 BRAIN STEM—MESENCEPHALON (MIDBRAIN)
Text pages 345-347

1. The midbrain connects structures of the _____ and the _____.

   1. _____

2. The midbrain is part of the brain _____.

   2. _____

Description/Structures	Cerebral Peduncles	Tectum
lateroventral portion	✔	
dorsal portion		✔
inferior and superior colliculi (corpora quadrigemina or quadrate bodies)		✔
tegmentum and crus cerebri separated by substantia nigra	✔	
important for integration of auditory and visual information		✔
large efferent (motor) tracts	✔	
important reflex centers		✔
gray matter surrounding cerebral aqueduct contains nuclei of oculomotor and trochlear nerves (C III and C IV)	✔	
red nuclei of the extrapyramidal tract	✔	

Note: If you remember that the eyes are superior (in location) to the ears, you will quickly associate the superior colliculi with vision and the inferior colliculi with audition.

No. 5-13 BRAIN STEM—MYELENCEPHALON: MEDULLA OBLONGATA
Text pages 347-348

1. The medulla oblongata is

   1.

   a. the inferior portion of the brain _____.

   a. _____

   b. continuous with the _____ inferiorly.

   b. _____

   c. continuous with the _____ superiorly.

   c. _____

2. In the medulla, pyramids are formed by a major *ascending / descending* pathway. It is a major *motor / sensory* pathway.

   Its fibers are *efferent / afferent*.

   2. _____

   _____

   _____

3. Over half of the fibers in the pyramids decussate at the anterior juncture of the medulla and the _____.

   3. _____

4. The fibers which decussate continue downward as the _____ pyramidal tract.

   4. _____

No. 5-13 cont'd

5. The fibers which descend uncrossed form the
    _____ pyramidal tract.

6. The inferior olives provide a relay station for proprio-
    ceptive impulses destined for the _____.

7. The fasciculus gracilis and the fasciulus cuneatus
    are *ascending / descending* tracts.

    They are *motor / sensory* tracts, particularly important
    for _____.

8. Posteriorly and superiorly, a pair of stalks
    carry fibers from the spinal cord and medulla
    to the *cerebellum / cerebrum*. These stalks are
    called the inferior _____ peduncles.

9. Three longitudinal columns of nuclei occupying the
    central core of the brain stem make up the
    _____ formation. These nuclei receive
    fibers from and contribute fibers to the _____
    nerves.

10. Regulatory centers in the medulla also influence
    respiration / reproduction / circulation / digestion.

5. _____

6. _____

7. _____

   _____

   _____

8. _____

   _____

9. _____

   _____

10. _____

    _____

Note: The term *bulbar* generally refers to the medulla oblongata.

Quote: "*progressive bulbar palsy*, progressive paralysis and atrophy of the muscles of the lips, tongue, mouth, pharynx, and larynx due to lesions of the motor nuclei of the lower brain stem. It is a chronic, generally fatal disease with onset usually in late adult years, but also earlier in patients with amyotrophic lateral sclerosis, syringobulbia, or multiple sclerosis."

"*pseudobulbar paralysis*, spastic weakness of the muscles innervated by the cranial nerves . . . due to bilateral lesions of the corticospinal tract; symptoms include dysphagia, dysarthria, and spastic facial jerks, sometimes accompanied by uncontrolled weeping or laughing and Cheyne-Stokes respiration."

Dorland's Illustrated Medical Dictionary, 1994

FIGURE 5.9 SOME DETAILS OF THE BRAIN STEM.

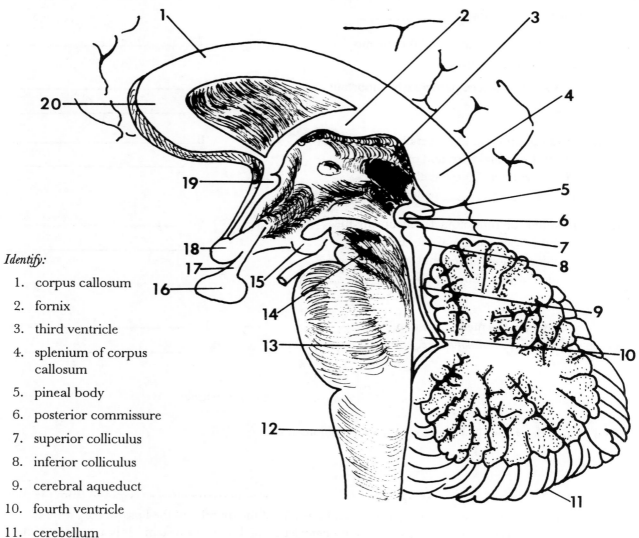

Identify:

1. corpus callosum
2. fornix
3. third ventricle
4. splenium of corpus callosum
5. pineal body
6. posterior commissure
7. superior colliculus
8. inferior colliculus
9. cerebral aqueduct
10. fourth ventricle
11. cerebellum
12. medulla oblongata
13. pons
14. mesencephalon
15. mammillary body
16. pituitary
17. infundibulum
18. optic chiasm
19. anterior commissure
20. genu of corpus callosum

1. The pons is a part of the brain _____.

2. The pons is immediately superior to the _____ and projects *anteriorly / posteriorly*.

3. The rounded, anteriorly directed portion of the pons consists of interlaced transverse fibers which collect laterally to form the middle cerebellar _____.
    These fibers function in part as a commissure, joining the two halves of the _____.

4. Behind this frontal yoke of fibers, the pons is continuous below with the _____, and above with the *cerebellar / cerebral* peduncles.

5. The central portion of the pons is composed of three longitudinal columns of nuclei, continuations of those found in the _____. These columns comprise the _____ formation..

6. The pontine nuclei relay impulses from the cerebral hemispheres to the _____.

1. _____

2. _____
    _____

3. _____
    _____

4. _____
    _____

5. _____
    _____

6. _____

Word Association: A pope is sometimes called a <u>pont</u>iff, one who establishes a *bridge* between his followers and God. In prosthodontics a <u>pont</u>ic is the portion of a *bridge* which replaces a missing tooth.

Note: The pons forms a *bridge* carrying afferent impulses from the spinal cord to the cerebellum and relaying impulses from the medulla oblongata to higher cortical centers. Its transmits integrated information about locomotion and body posture, e.g., it tells the body at an unconscious level whether or not it is in danger of falling down.

Quote: "The most common knockout blow is one that makes contact with the mandible. Such a blow twists and distorts the brain stem and overwhelms the reticular activating system (RAS) by sending a sudden volley of nerve impulses to the brain, resulting in *unconsciousness*."

Gerald J. Tortora, 1989

FIGURE 5.10 THE CEREBELLUM

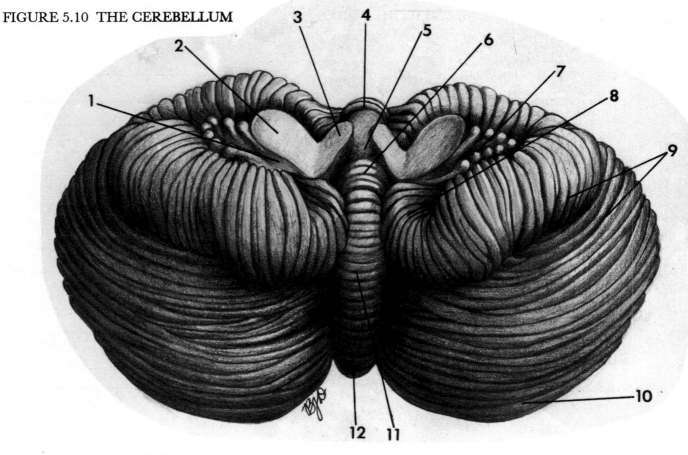

Identify:
1. inferior cerebellar peduncle
2. middle cerebellar peduncle
3. superior cerebellar peduncle
4. central lobule
5. fourth ventricle
6. nodulus
7. flocculus
8. tonsil
9. biventral lobule
10. inferior semilunar lobule
11. uvula
12. pyramis

No. 5-15 BRAIN—METENCEPHALON: CEREBELLUM
Text pages 349–354

1. The cerebellum makes up the greater part of the _____ brain.

   1. _____

2. The shape of the cerebellum is somewhat _____, with a pronounced medial *expansion / constriction*.

   2. _____
      _____

3. The many transversely directed cortical ridges are called folia _____.

   3. _____

4. Just deep to the cortex lies the branchlike white matter, the _____ vitae.

   4. _____

5. The narrow median portion, the _____, is partially covered by the cerebellar _____.

   5. _____

6. As you can see, there is a structural resemblance between the cerebellum and the _____.

   6. _____

Description/Strucctures	Archicerebellum	Paleocerebellum	Neocerebellum
old; second to develop embryologically		✔	
new			✔
first to develop embryologically	✔		
flocculus and nodulus	✔		
highly developed in humans			✔
primarily functions to control limbs		✔	
vestibular function; body position; equilibrium	✔		

Description Cortical layers:	Molecular	Purkinje	Granular
middle layer		✔	
outermost layer	✔		
innermost layer			✔
the final pathway out of the cortex (inhibitory output)		✔	
contains the very smallest neurons; densely packed			✔
contains basket, Golgi and stellate cells and also nerve fibers	✔		
its cells are excited by afferent fibers entering cerebellum via cerebellar peduncles			✔

Description Cerebellar Peduncles:	Superior	Middle	Inferior
also called brachium conjunctivum	✔		
also called restiform body			✔
also called brachium pontis		✔	
largest		✔	
carries olivocerebellar, dorsospinocerebellar, and vestibulocerebellar fibers			✔
carries fibers from nuclei in pons to contralateral neocerebellum		✔	
carries dentaterubral, ventral spinocerebellar, and uncinate fasciculus fibers	✔		
carries fibers between brain stem and cerebellum	✔	✔	✔

Note: The structural arrangement of cells in the cerebellar cortex allows for *avalanche conduction* of nerve impulses. The terminals of one neuron contact several nerve cell bodies, thus causing widespread discharge.

Quote: "To cause serious and continuous dysfunction of the cerebellum, the cerebellar lesion must usually involve the deep cerebellar nuclei . . . as well as the cerebellar cortex."

A. C. Guyton, 1981

FIGURE 5.11 THE STRUCTURE OF THE CEREBELLUM

One difficulty in understanding the structure of the cerebellum, besides inconsistent vocabulary, is the way in which it is projected in most illustrations.

To restore proper spatial relationships, fold this page under at the horizontal fissure so the flocculus, nodule, pyramid, and tuber are on the underside.

The flocculus and nodulus are now located beneath the lingula at the anterior cerebellum and the tuber is at the posterior cerebellum.

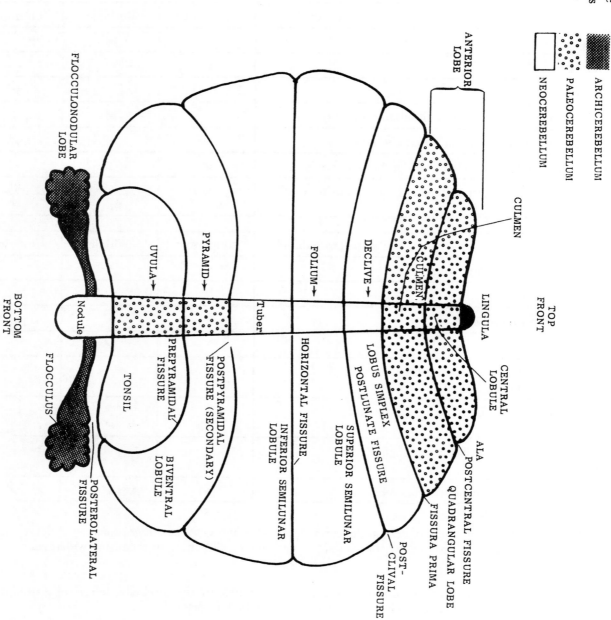

1. The cerebellum acts on impulses initiated *in / outside* the cerebellum.

1. _____

2. The cerebellum insures smooth and coordinated *voluntary / involuntary / voluntary and involuntary* movements.

2. _____

3. The cerebellum compares what the body is doing with "decisions" being made by the _____ cortex, and provides the necessary _____.

3. _____

4. The activity of structures in the right side of the cerebellum affects the *right / left* side of the body. Cerebellar function is therefore said to be _____.

4. _____
_____

5. Why may cerebellar symptoms result from lesions outside the cerebellum?

5. _____
_____
_____

6. The cerebellum has a relatively *high / low* compensatory potential probably due in part to the large neocerebellar _____.

6. _____
_____

Function	Archicerebellum	Paleocerebellum	Neocerebellum
control of synergistic movement			✔
maintenance of equilibrium	✔		
control of gait	✔		
largely determined by animal experiments		✔	
control of antigravity musculature		✔	
control of range of voluntary motor acts			✔

Cerebellar Symptoms	Description
asthenia	weakness, easily tired
ataxia	uncoordinated, irregular muscle action
disequilibrium	unsteadiness, tendency to fall
dysmetria	difficulty measuring distance of muscle movement
dysdiadochokinesia	inability to perform rapidly changing movements
dysarthria	neurologically-based impairment of muscular control for speech
movement decomposition	complex movement becomes series of disconnected simpler movements
cerebellar nystagmus	rapid, involuntary movement of eyeball
positional:	posture of the body may influence nystagmus
directional:	direction of gaze may influence nystagmus
intention tremor	involuntary trembling; intensified during voluntary movement
rebound	overmovement due to lack of antagonistic muscle movement
past pointing	symptom of dysmetria; moving far beyond intended target
hypotonia	decreased muscle tone; lessened resistance to stretching
scanning speech	slurred; inappropriate blending of sounds

No. 5-17 THE SPINAL CORD (VOCABULARY)
Text pages 356–360

cauda equina   horns
central canal   intervertebral foramen
filum terminale   laminae
funiculi   nuclei

1. cavity of the spinal cord

   1. _____

2. spinal nerves in inferior portion of spinal column; like a horse's tail

   2. _____

3. prolongation of inferior tip of spinal cord

   3. _____

4. columns of white matter within the spinal cord

   4. _____

5. columns of aggregates of nerve cell bodies within the spinal cord; having a specific function (3)

   5. _____
      _____
      _____

6. exit for a spinal nerve

   6. _____

No. 5-18 THE SPINAL CORD
Text pages 356–360

1. In an adult the spinal cord extends from the first _____ vertebra to the first _____ vertebra.

   1. _____
      _____

2. In the fetus the spinal nerves exit the spinal cord at approximately _____ angles. As the spinal column grows, the lower spinal nerves exit at increasingly *obtuse / acute* angles.

   2. _____
      _____

3. Enlargements of the spinal cord in the cervical and lumbar regions are due to increased nerve supply needed for the _____.

   3. _____

4. Approximately how many pairs of spinal nerves emerge from the spinal cord?

   4. _____

5. The spinal cord is incompletely divided into right and left halves by an anterior and posterior median _____.

   5. _____

6. The central core of gray matter is said to resemble the letter _____ or the shape of a/an _____.

   6. _____
      _____

7. Tracts within the white matter may exhibit topographical distribution. This may provide useful _____ information.

   7. _____

Description/Function Gray columns:	Ventral (anterior horn)	Dorsal (posterior horn)	Lateral
present in upper cervical, thoracic, and midsacral regions			✔
contains many motor neurons	✔		
contains many sensory and internuncial neurons		✔	
contains many autonomic neurons			✔
generally associated with motor functions	✔		✔
generally associated with receptor and coordinating functions		✔	
extends full length of spinal cord	✔	✔	
joined by transverse commissure	✔	✔	

Tracts/Fibers Funiculi (white):	Ventral (anterior)	Dorsal (posterior)	Lateral
ascending tracts supplying visceral/proprioceptive data to subcortical motor centers	✔	✔	✔
descending tracts from higher motor centers	✔	✔	✔
short intersegmental fibers that mediate reflexes	✔	✔	✔
pyramidal (corticospinal) tract (crossed)			✔
rubrospinal (an extrapyramidal tract)			✔
fasciculus gracilis; fasciculus cuneatus		✔	

Clinical Signs/Description Injury to Motor Neurons:	Upper	Lower
damage in corticospinal tract above the level of the synapse	✔	
damage in corticospinal tract below the level of the synapse		✔
flaccid paralysis		✔
spastic paralysis	✔	
increased deep tendon reflexes	✔	
loss of deep tendon reflexes		✔
no muscle atrophy	✔	
muscle atrophy		✔
paralysis below level of injury	✔	✔
fasciculations and fibrillations		✔
no fasciculations and fibrillations	✔	
anesthesia below level of injury	✔	✔

Summary of the major tracts of the spinal cord. Table 5-1, text page 359.

1. Cells of the cerebral cortex:

    a. The two predominant types are _____ and_____.

    b. Giant pyramidal cells of Betz, contained in the fifth layer, initiate *motor / sensory* impulses.

    c. The most superficial layer is made up of _____ neurons.

2. What are some of the methods that have been used to explore the relationship between function and cytoarchitectonics (cellular architecture) and myeloarchitectonics (nerve fiber distribution)?

    _____

    _____

3. Brodmann developed an architectonic chart (_____ map) based on cytoarchitecture.

4. The relationship between the amount of cortical representation and

    a. structure size is usually *direct / inverse.*

    b. extent of innervation is usually *direct / inverse.*

    c. body area is graphically represented by a/an _____.

5. Motor areas:

    a. Stimulation of one side of the motor cortex usually produces movement on the *contralateral / ipsilateral* side.

    b. Regions of the body more likely to be bilaterally controlled are located *medially / laterally.*

    c. The premotor area is like the motor area, but has no giant _____ cells. It contains fibers from the *pyramidal / extrapyramidal* tract.

    d. The motor speech area is called _____'s area.

6. Sensory areas:

    a. Topographical projections of the sensory and motor areas are very *similar / different.*

    b. The somatic sensory area and the somatic motor area are separated by the central fissure. The sensory area is *anterior / posterior* to the fissure.

    c. The auditory sensory area is located on the *temporal / occipital* gyri.

    d. An auditory association area relating past experiences and present sensations is called _____'s area.

1.
    a. _____
       _____
    b. _____
    c. _____

2.
    _____
    _____

3. _____

4.
    a. _____
    b. _____
    c. _____

5.
    a. _____
    b. _____
    c. _____
       _____
    d. _____

6.
    a. _____
    b. _____
    c. _____
    d. _____

e. The pathway between the auditory association areas and the motor speech area is the _____ fasciculus.

e. _____

f. The striate area is the cortical center for *hearing / vision*.

f. _____

g. The parastriate area is a visual _____ area.

g. _____

h. The vestibular area is involved in the recognition of _____ and _____.

h. _____

_____

7. Regions that inhibit cortical activity are called _____ regions.

7. _____

8. Aphasia:

8.

a. Aphasia is usually the result of a lesion in the *right / left* hemisphere.

a. _____

b. An aphasic who appears to be talking nonsense probably has a lesion in _____'s area.

b. _____

c. An aphasic who comprehends speech and can communicate nonverbally probably has a lesion in _____'s area.

c. _____

d. The arcuate fasciulus transmits information from _____'s area to _____'s area.

d. _____

_____

e. Recovery is usually more complete in *children / adults*.

e. _____

Symptoms/Etiology	Apraxia	Dysarthria
language disorder		
speech disorder	✔	✔
lesion in association cortex (dominant)	✔	
lesion in central or peripheral nervous system		✔
lesion may involve upper or lower motor neurons		✔
no paralysis	✔	
hypertonicity (upper motor neuron involvement)		✔
hypotonicity (lower motor neuron involvement)		✔
impaired motor programming	✔	
slow, labored, imprecise articulation		✔
speech mechanism itself is not impaired, but voluntary control is lacking	✔	

Quote: "Because bilingual speakers who develop aphasia usually experience difficulties in both languages, it appears that the basic brain organization of the two languages is the same. An exception may occur in the cases where one of the languages is based on an alphabet and the other is based on ideographs, symbols that represent objects or ideas rather than sounds. Chinese is an ideographic language, for instance. Differences in reading ability after brain damage suggest that the different kinds of language are organized differently in the brain."

F. E. Bloom and A. Lazerson, 1988

No. 5-20 HEMISPHERIC DOMINANCE
Text pages 363–366

Description/Characteristics	Hemisphere:	Right	Left
dominant in 90% right-handed population			✔
dominant in 64% left-handed population			✔
dominant in 20% left-handed population		✔	
dominant in 60% ambidextrous population			✔
dominant in 30% ambidextrous population		✔	✔
usually functions in touch perception, spatial concepts, memory, nonverbal language		✔	
usually functions in verbal language, calculation, memory			✔
important in holistic and creative processes, art, music		✔	
usually dominant in a one-year-old child			
frontal lobe usually larger and more convoluted			✔

Quote: "Connectionist, or neural network, theories try to explain various behaviors or mental feats by drawing analogies with the learning exhibited by interconnected processing units in a computer system. The mathematical strength of connections among these units gradually changes as the system learns to solve a particular problem. The computer network eventually assumes a stable pattern of activity that successfully tackles both the familiar problem and related new challenges. . . . Connectionist systems are edging toward realistic simulations of how children may attain grammatical milestones through learning."

Bruce Bower, 1997

No. 5-21 THE RETICULAR FORMATION
Text pages 366–367

1. Awake State:
   a. Multisensory information about the internal and external environment is carried by ascending tracts to nuclei in the _____ formation.
   b. This information is transmitted to "nonspecific" nuclei in the _____, and projected to large areas of the _____.
   c. This increased awareness of one's self and one's environment contributes to _____
2. Sleep State:
   a. Sleep occurs when awareness of self and and environment _____.
   b. Dreaming occurs during _____ sleep.
3. Input to the reticular formation can excite or suppress *small / large* numbers of other neurons.

1.
   a. _____
   b. _____
      _____
   c. _____
2.
   a. _____
   b. _____
3. _____

FIGURE 5.12 SCHEMATIC OF THE NERVOUS SYSTEM.
Based on a woodcut by Andreas Vesalius (1514-64).

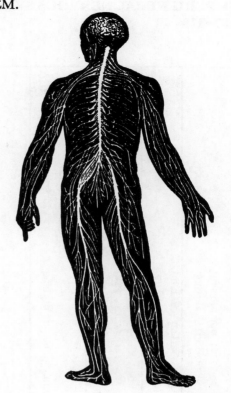

No. 5-22 THE PERIPHERAL NERVOUS SYSTEM
Text pages 367 *Review No. 5-4, page 182.*

Note: The term somatic pertains to the body, especially the voluntary muscles and skeletal framework.

Types of nerve fibers in the peripheral nervous system:

general somatic afferent
special somatic afferent
somatic efferent
general visceral afferent
special visceral afferent
general visceral efferent
special visceral efferent

Distribution/Function	General	Special	Somatic	Visceral	Afferent	Efferent
in some cranial and all spinal nerves; supply striated muscles			✔			✔
in optic and auditory nerves		✔	✔		✔	
only in cranial nerves; name is misleading as they supply striated muscles of larynx, pharynx, soft palate, mastication, facial expression		✔		✔		✔
in all spinal and some cranial nerves; conduct impulses from integument, muscles, connective tissue to CNS	✔		✔		✔	
in cranial and spinal nerves; serve viscera of neck, thorax, abdomen, pelvis, also blood vessels and glands	✔			✔	✔	
only in olfactory, glossopharyngeal, and vagus nerves; sense of taste and smell		✔		✔	✔	
in cranial and spinal nerves and peripheral ganglia of ANS; supply smooth muscle and glands	✔			✔		✔

Cranial Nerves:

1. Motor cranial nerves arise from nuclei *within / outside* the brain stem.

2. Sensory cranial nerves arise from ganglia *within / outside* the brain stem.

1. _____

2. _____

No.	Mnemonic Device	Name of Nerve	Function	Name Reflects:	Mnemonic Device	Type (sensory, motor, both)	Symptoms of Lesions
I	On	olfactory	smell	function	Some	sensory	loss of sense of smell
II	old	optic	vision	function	say	sensory	*nerve* - varying losses of vision in one eye; *chiasm* - bilateral loss of lateral vision; *tract* - loss of right/left half of visual field
III	Olympus'	oculomotor	visual convergence, pupil size/lens shape	function	Marilyn	motor	interrupts conjugate/convergent eye movement, drooping eyelids, double vision, dilated pupils
IV	towering	trochlear	visual tracking (downward/outward)	structure (pulley)	Monroe	motor	double vision when looking downward/outward
V	top	trigeminal	sensations to facial regions, controls mastication	structure (three twins)	but	both	facial numbness, paralysis of muscles of mastication, loss of corneal reflex, loss of muscle tone in floor of mouth, increased sensitivity to sound, intense facial pain
VI	a	abducens	lateral eye movement	function (separate)	my	motor	internal strabismus (eye pulls to nasal side), double vision
VII	Finn	facial	sensations to tongue, soft palate, facial expression, taste	distribution	brother	both	Bell's palsy; paralysis of contralateral facial muscles, stapedius muscle (middle ear); loss/lack of tears, saliva, taste; sensitive to low frequency sounds
VIII	and	acoustic (vestibulocochlear)	hearing and equilibrium	function	says	sensory	*cochlear (acoustic)* - loss of hearing (ipsilateral) *neuroma* - facial pain/numbness, headache, tinnitus *vestibular* - vertigo, nystagmus, unsteadiness
IX	German	glossopharyngeal	swallowing, salivation, taste	distribution	Bridget	both	difficulty initiating swallow, loss of taste/sensation back ⅓ of tongue, unilateral loss of gag reflex, deviation of uvula, tachycardia
X	viewed	vagus	sensations to/control of pharynx, larynx, viscera	structure (wander)	Bardot	both	paralysis of soft palate (nasality), dysphagia, deviation of uvula, aphonia, breathiness, rough voice quality
XI	a	accessory	activate sternocleido-mastoid and trapezius	function	Mmm	motor	paralysis of sternocleidomastoid, inability to turn head, shrug shoulders, raise arms above shoulders; voice problems
XII	hop.	hypoglossal	control of tongue movement	distribution	Mmm.	motor	tongue deviation, fasciculation; eventual atrophy of affected side of tongue; articulation could be affected

Function/Distribution	Trigeminal V	Facial VII	Acoustic VIII	Glossopharyngeal IX	Vagus X	Accessory XI	Hypoglossal XII
both motor and sensory	V	VII		IX	X		
only motor						XI	XII
only sensory			VIII				
most extensively distributed					X		
largest	V						
cochlear branch (sensory nerve for hearing)			VIII				
origin of ophthalmic, mandibular and lingual nerves	V						
courses through neck and thorax; extends into abdomen					X		
parasympathetic fibers innervate mucous membrane of middle ear and auditory (Eustachian) tube				IX			
communicates with many other cranial nerves		VII			X		
motor fibers innervate tongue and pharynx				IX			
vestibular branch (sensory nerve of equilibrium)			VIII				
some autonomic innervation of pharynx, palate, nasal cavity, and paranasal sinuses		VII					
visceral sensory innervation of taste buds on posterior portion of tongue				IX			
origin of motor fibers which enter vagus and emerge as recurrent laryngeal nerve						XI	
sensory innervation of taste buds on anterior two-thirds of tongue		VII					
motor innervation of muscles of soft palate and of mastication	V						
motor innervation of intrinsic and extrinsic muscles of the tongue							XII
motor innervation of muscles of facial expression and the stapedius muscle		VII					
sensory innervation of superficial and deep structures of the face, mouth, and lower jaw	V						

FIGURE 5.13 SCHEMATIC BRAIN STEM SHOWING CEREBRAL PEDUNCLES AND THEIR RELATIONSHIP TO THE PONS AND OPTIC TRACT. THE EMERGENCE OF THE CRANIAL NERVES IS ALSO SHOWN.

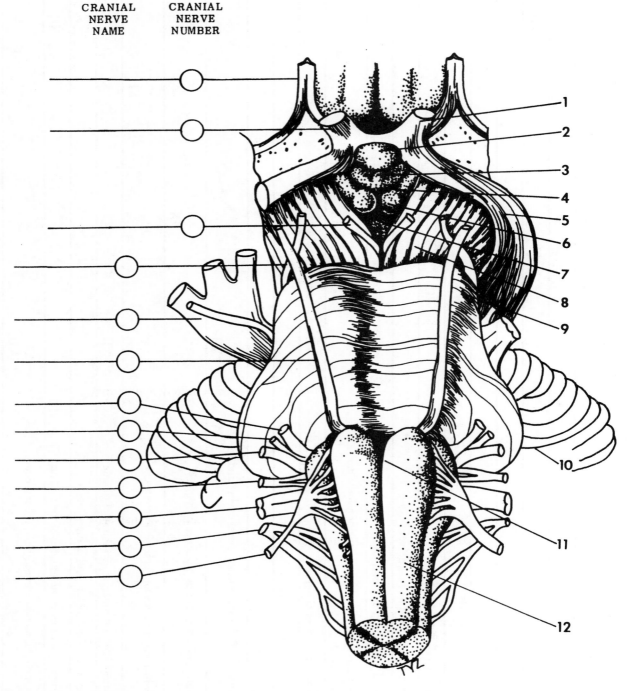

CRANIAL NERVE NAME

CRANIAL NERVE NUMBER

Identify:

1. optic chiasma
2. hypophysis
3. infundibulum
4. tuber cinereum
5. optic tract
6. mammillary body

7. posterior perforated substance
8. cerebral peduncle
9. pons
10. cerebellum
11. pyramids (of medulla oblongata)
12. medulla oblongata

1. Dorsal (afferent) and ventral (efferent) roots arise from the *gray / white* matter within the spinal cord.

   1. _____

2. Dorsal spinal roots carry nerve processes of *motor / sensory* neurons arising from the *central / peripheral* nervous system.

   2. _____

     _____

3. Ventral spinal roots carry nerve processes of the *motor / sensory* neurons arising from the *central / peripheral* nervous system.

   3. _____

     _____

4. Near or within each intervertebral foramen is a dorsal root enlargement called the _____ ganglion.

   4. _____

5. Just beyond the spinal ganglion the dorsal and ventral roots join to form a/an _____ nerve.

   5. _____

6. The cutaneous area served by one sensory nerve is called a/an _____. The emergence of the spinal roots at somewhat regular intervals to supply rather specific parts of the body is sometimes called _____ innervation.

   6. _____

     _____

7. The spinal nerve exits through the _____ foramen, and immediately divides into anterior and posterior _____. They carry *sensory / motor / sensory and motor* fibers.

   7. _____

     _____

     _____

8. Postural muscles are supplied by the _____ rami.

   8. _____

9. The anterior rami of the first four cervical nerves branch and interlace to form the cervical _____. It communicates with some of the _____ nerves.

   9. _____

     _____

10. The nerve arising from the cervical plexus and innervating the diaphragm is the _____ nerve.

  10. _____

11. The communicating branches of the lower four cervical and first thoracic nerves form the _____ plexus. It supplies the _____ and _____.

  11. _____

     _____

     _____

12. The trunk and lower limbs are supplied by the _____ plexus.

  12. _____

1. The autonomic nervous system supplies the glands and
   smooth / striated / cardiac muscle. (two answers)

   1. _____

      _____

2. The autonomic nervous system or visceral efferent system
   is also known as the *voluntary / involuntary* system.

   2. _____

3. An impulse from the central nervous system to a visceral
   effector is carried by *one, two* neuron/s, while an
   impulse to a skeletal effector is carried by _____
   neuron/s.

   3. _____

      _____

4. Most organs are innervated by the *sympathetic division /
   the parasympathetic division / both divisions.*

   4. _____

5. Which division mobilizes the body in an emergency?

   5. _____

6. Which division conserves the body's resources?

   This division is sometimes called the vegetative division.

   6. _____

7. Explain why the preganglionic fibers of the sympathetic
   division are much shorter than those of the
   parasympathetic division.

   7.

   _____

   _____

   _____

8. Parasympathetic fibers originating in the

   8.

   a. midbrain are called *bulbar / tectal* autonomics.

      a. _____

   b. medulla and pons are called *bulbar / sacral* autonomics.

      b. _____

   c. spinal cord are called *sacral / tectal* autonomics.

      c. _____

9. Autonomic nerves carrying visceral afferent fibers
   are found in cranial nerves VII, IX, and X. Name
   those nerves.

   9. _____

      _____

      _____

Study Fig. 5-62 and Fig. 5-63, text pages 378–379.

Note: To help recall the functional divisions of the autonomic nervous system remember that you are
sympathetic to someone who has an *emergency*.

Studying the effects of drugs on the neural system has contributed to an understanding of the
autonomic nervous system.

Description Neurons:	Unipolar	Bipolar	Multipolar
derived from neuroblasts	✔	✔	✔
derived from glioblasts			
many dendrites and one axon			✔
seems to have one axon with two limbs	✔		
one axon and one dendrite		✔	
found in cerebral and cerebellar cortexes	✔		✔
associated with the senses	✔		
motor neuron having a long axon extending from the gray matter of the spinal cord to the skeletal muscle			✔

Description Neuroglial cells:	Ependymal	Astrocytes	Oligodendrocytes
derived from neuroblasts			
derived from glioblasts	✔	✔	✔
similar in function to satellite cells		✔	
have same function in CNS that Schwann cells have in peripheral nervous system			✔
secrete/absorb cerebrospinal fluid	✔		
increase conduction velocity through myelinization			✔
intermediary between neurons and blood capillaries		✔	
barrier between impurities in central canal/ventricular fluids and brain tissue	✔		

1. Axons

   a. are nerve _____.

   b. conduct impulses *toward / away from* the cell body.

   c. conduct nerve impulses to another neuron, a/an _____ cell, or a/an _____.

   d. and cell body junctions form _____ hillocks. Nerve impulses originating at these junctions are *efferent / afferent*. These junctions are *always / never / sometimes* myelinated.

   e. usually branch near their *cell body / termination*.

   f. branch out as telodendria, collectively known as end _____. At the end of each branch are synaptic _____.

   g. carry enzymes and chemicals needed to synthesize a/an _____ substance.

1.

   a. _____

   b. _____

   c. _____

      _____

   d. _____

      _____

      _____

   e. _____

   f. _____

      _____

   g. _____

No. 5-27 cont'd

    h. contain supportive structures called _____.

    i. contain mitochondria, the synthesizers of ATP (adenosine tryphosphate). ATP helps produce the nerve _____ potential.

2. Dendrites conduct impulses *toward / away from* the cell body.

3. Neurilemma (Schwann) cells

    a. form myelin in the _____ nervous system.

    b. are possessed by *myelinated / unmyelinated / all* axons in that system.

    c. play an important role in nerve fiber _____.

4. The junctions between neuroglial cells are called the nodes of _____.

5. Nerve fibers of the central nervous system *always / never / sometimes* regenerate.

6. A severed peripheral nerve fiber

    a. will *never / always / sometimes* regenerate.

    b. begins to degenerate *distally / proximally*. This is called *retrograde / Wallerian* degeneration.

    c. degenerates, but the neurilemmal sheath forms _____ tissue.

    d. may also undergo changes proximally, in a process called _____ degeneration.

    e. may continue to degenerate, eventually die, and be consumed by _____ cells.

    f. may begin to grow, sending sprouts through _____ tissue barrier. Those that do not cross may form a painful _____.

7. A crushed peripheral nerve

    a. will usually recover because the sprouts are confined within the neurilemmal _____.

    b. will usually recover within *3–6 / 6–12 / 12–18* months, but reinnervation will not restore a/an _____ muscle.

Description Tissue:	Endoneurium	Perineurium	Epineurium
connective tissue	✔	✔	✔
covers nerve processes	✔		
covers bundles of nerve processes		✔	
covers nerve trunks			✔

No. 5-28 NEURON EXCITATION AND CONDUCTION: RESTING MEMBRANE POTENTIAL
Text page 385–387

1. Electrical potentials can be developed by *all, some* living tissues, both plant and animal.

    1. _____

2. Action potentials can be produced only by _____, _____, and _____.

    2. _____

    _____

    _____

3. In a state of rest, any two points on the surface of a cell membrane exhibit *the same / different* electrical potentials.

    3. _____

4. Electrical potentials exhibited on the surface of the cell membrane and those of the cytoplasm within the cell membrane are *the same / different.*

    4. _____

5. In an atom the negatively and positively charged electrons are *balanced / unbalanced.*

    5. _____

6. An atom that has acquired an electrical charge by gaining or losing an orbital electron is called a/an _____.

    6. _____

7.

Characteristics/Action	Pairs of Ions:	+ +	− −	+ −
a. attract each other				
b. repel each other				
c. separation generates an electrical force				
d. poles of a flashlight cell				

8. Potential force

    8.

    a. varies directly with *number of / distance between* ions.

    a. _____

    b. varies inversely with *number of / distance between* ions.

    b. _____

    c. is measured in units called *ohms / volts.*

    c. _____

9. The flow of ions

    9.

    a. requires that the poles be connected by an electrical *insulator / conductor.*

    a. _____

    b. is called electrical _____.

    b. _____

    c. is measured in units called *amperes / volts.*

    c. _____

    d. is affected by the resistance to electrical flow, as measured in units called *watts / ohms.*

    d. _____

10. Resting membrane potential

10.

    a. is a bioelectrical potential force that can be measured in *volts / ohms*.

a. _____

    b. is due to polarization of the cell _____. More positive ions are concentrated *inside / outside* the cell.

b. _____

_____

    c. is dependent on the characteristic of the plasma membrane called selective _____.

c. _____

11.

Characteristics Ions:	Potassium (K⁺)	Sodium (Na⁺)	Chloride (Cl⁻)
a. contain water; too large to readily diffuse through membrane			
b. small; readily diffuse through membrane			

12. The sodium potassium pump

12.

    a. is a property of excitable cell _____.

a. _____

    b. ejects *sodium / potassium* ions from the cell.

b. _____

    c. retrieves *sodium / potassium* ions that were lost.

c. _____

    d. maintains _____ potential.

d. _____

1. The stimulus most often used as a source of artificial stimulation for neurons is _____.

   1. _____

2. Stimulation of a neuron:

   2.

   a. Initial stimulation results in *polarization / depolarization* of the cell membrane.

   a. _____

   b. Resting membrane potential *increases / decreases*.

   b. _____

   c. If stimulation increases, polarity of the cell membrane will be *the same / reversed*.

   c. _____

   d. The cell membrane enters a _____ phase.

   d. _____

   e. Resting membrane potential is restored due to action of the _____.

   e. _____

3. Stimulation requirements:

   3.

   a. The minimum strength of electrical current required to stimulate a nerve or fiber is called *rheobase / chronaxie*.

   a. _____

   b. The time required for a current twice the minimal strength to stimulate a nerve is called _____ time or _____.

   b. _____
   _____

   c. These measures allow you to determine the relative _____ of different nerves and fibers.

   c. _____

4. Potentials:

   4.

   a. Electric changes which accompany repolarization are called _____-potentials.

   a. _____

   b. The most conspicuous electrical change in a recording of a nerve impulse, the sharp reversal of potential due to cell depolarization, is called _____ potential.

   b. _____

   c. Together these two potentials represent the nerve impulse or the nerve-_____ potential.

   c. _____

5. If the strength of a stimulus is greatly increased

   5.

   a. the character of the individual impulses will change.

   a. T   F

   b. the frequency of the nerve impulses will increase.

   b. T   F

   c. more nerve fibers will be stimulated.

   c. T   F

   Would a, b, or c contradict the all-or-none principle?

   _____

6. If a motor nerve is stimulated by a succession of stimuli the resulting contractions may become fused, resulting in a continued contractile state called _____.

   6. _____

7. Conduction in myelinated and unmyelinated fibers:

   7.

  a. The relationship between nerve fiber diameter and conduction velocity is *direct / inverse*.

     a. _____

  b. Given two equal nerve fibers, one myelinated and one unmyelinated, which should conduct more rapidly?

     b. _____

  c. In myelinated fibers ions can flow outward only at the nodes of _____.

     c. _____

  d. Continous conduction is characteristic of _____ fibers.

     d. _____

  e. Saltatory conduction is characteristic of _____ fibers. Describe it.

     e. _____

_____

_____

Description	Refractory Period		Phase	
	Absolute	Relative	Supernormal	Subnormal
modifications of response characteristics of a nerve fiber over a period of time	✔	✔	✔	✔
nerve fiber is more excitable than in resting state			✔	
time occupied by spike potential; completely depolarized cell is unable to respond to another stimulus	✔			
corresponds to duration of positive after-potential				✔
corresponds to duration of negative after-potential			✔	
stronger stimuli required to initiate responses; those responses will be of reduced magnitude		✔		
included in *total* refractory period	✔	✔		
toilet flushed / tank empty	✔			
toilet flushed / tank half full		✔		

Review the Characteristics of Action Potentials, text page 390.

1. The functional connections between neurons are known as neuronal junctions or _synapses_. Is there cytoplasmic continuity of neurons at these junctions? No

1. _____

2. The minute space between the terminal button and the indentation of the membrane of the succeeding cell is called a/an _synaptic_ cleft.

2. _____

3. A single cell may synapse with as many as 1,000 / 50,000 / (100,000) other cells.

3. _____

4. Acetycholine

   a. is a _neurotransmitter_

   b. is released at the _terminal_ buttons.

   c. stimulates the _postsynaptic_ cell.

   d. is destroyed by acetylcholinesterase, a/an (enzyme) / neurotransmitter.

4.

   a. _____

   b. _____

   c. _____

   d. _____

5. The synapse is the part of the nervous system most sensitive to lack of _oxygen_.

5. _____

6. Inadequate stimuli may produce synaptic transmission if impulses arrive at very close intervals. This is called spatial / (temporal) summation.

6. _____

7. Inadequate stimuli may produce synaptic transmission if impulses from a number of axons arrive simultaneously. This is called (spatial) / temporal summation.

7. _____

Neural Synapse:	Neuromuscular Synapse:
membrane of synaptic region	motor end plate
synaptic cleft	synaptic cleft
boutons	boutons
liberation of acetycholine	liberation of acetycholine
nerve action potential (chemically produced)	muscle action potential (chemically produced)

8. Stimulation of peripheral motor neurons results in muscle _contraction_

8. _____

9. The regulatory activity responsible for the relaxation of an antagonistic muscle is called nervous _inhibition_. For voluntary muscles, this type of regulation probably resides in the (central) / peripheral nervous system.

9. _____

Study Fig. 5-62, text page 379, and label structures in Fig. 5-72, text page 387.

Characteristics/Function	Mechanoreceptors	Thermoceptors	Nociceptors	Photosensreceptors	Chemoreceptors
responsible for taste and smell					chemo
detect temperature changes		therm			
respond to tissue damage			noci		
respond to mechanical pressure	mech				
detect light directed on the retina				photo	
respond to deformation of receptor and adjacent tissue	mech				
exteroceptors (respond to immediate environment; located or near the skin)	mech	therm			
kinesthetic receptors (in joint capsules and ligaments)	mech				
activation mediates sensation of pain			noci		
when stimulated will produce local electrical changes called receptor potentials	mech	therm	noci	photo	chemo
have no refractory period	mech	therm	noci	photo	chemo
olfactory receptors and taste buds					chemo
rods and cones of retina				photo	
transmit signals to central nervous system; ultimately reach conscious levels	mech	therm	noci	photo	chemo
relationship between form and function	mech	therm	noci	photo	chemo
characterized by adaptation (cessation or decrement of nerve-impulse frequency in spite of continuous stimulation)	mech	therm	noci	photo	chemo
Meissner's corpuscles, free nerve endings, expanded tip receptors, merkel's discs, hair end organs, Ruffini's end organ, Pacinian corpuscles, and spray endings	mech				

Note: Information of an acute nature is delivered to the central nervous system by large fibers that conduct impulses very rapidly. Chemical, thermal and electrical stimuli, when extremely strong, may arouse not only receptors specialized to respond to those stimuli, but also other receptors.

No. 5-32 MUSCLE AND TENDON RECEPTORS
Text pages 397–400

Description/Function Receptors:	Muscle Spindles	Golgi Tendon Organs
receptors operating on a subconscious level, generating no sensory awareness	✔	✔
respond to muscle fiber *length* and its rate of change	✔	
detect *tension* generated in tendon fiber as a consequence of muscle contraction		✔
information transmitted may terminate at motor control mechanisms at spinal level	✔	✔
information transmitted may terminate in cerebellum	✔	✔
responsible for reflexes associated with equilibrium, posture, and fine control of muscle movement	✔	✔
proprioceptors; usually near junctions of muscles and tendons		✔
initially respond with a burst of impulses, then maintain a slower but constant rate of discharge	(primary endings)	✔
stretch receptors	✔	
signals transmitted from this receptor to the internuncial neurons in spinal cord generate inhibitory reflex behavior		✔
composed of intrafusal muscle fibers surrounded by extrafusal muscle fibers	✔	
stimulated when their mid-regions are streched	✔	

No. 5-33 MUSCLE SPINDLES (STRETCH RECEPTORS)
Text pages 397–399

1. The relatively thick connective tissue capsule encasing the intrafusal fibers of a muscle spindle is expanded into a fluid-filled sac in the central or _____ region.

   1. ___*equatorial*___

2. Intrafusal fibers:

   2.

   a. are totally encased in the capsule.

      a. T (F)

   b. project from both ends of the capsule.

      b. (T) F

   c. project from only one end of the capsule.

      c. T (F)

   d. are parallel to the extrafusal fibers.

      d. (T) F

   e. join with adjacent extrafusal fibers at the midline.

      e. T (F)

   f. join with extrafusal fibers at their tapered ends.

      f. (T) F

   g. are fusiform or _____-shaped.

      g. ___*spindle*___

   h. are the contractile substance of muscle tissue.

      h. (T) F

Description Fibers:	Nuclear Chain	Nuclear Bag
have larger diameter		✔
nuclei distributed end to end along axis of fiber	✔	
characterized by clusters of nuclei, particularly in equatorial region of spindle		✔
density of striations greatest at polar extremes	✔	✔
contractile only at their ends or poles	✔	✔
receive large alpha type A afferent nerve fiber	✔	✔
type of intrafusal fiber found in muscle spindles	✔	✔

Characteristics Receptors:	Primary	Secondary
respond very quickly	✔	
respond very slowly		✔
transmit initial burst of impulses, followed by slower steady state impulses	✔	
react primarily to actual length of receptors		✔
annulospiral (spiral) endings formed by large alpha type A motor fiber winding around muscle fibers	✔	
flower spray endings which excite beta type A motor fibers		✔
stimulated when stretch of entire muscle belly stretches equatorial region of muscle spindle	✔	✔
stimulated when ends of intrafusal fibers contract, thereby stretching equatorial region of muscle spindle	✔	✔
transmit information about actual length or receptor	✔	

3. Gamma Efferent Control of the Sensitivity of Muscle Spindles

3.

a. The gamma efferent fibers supply the *intrafusal / extrafusal* fibers of the muscle spindle.

a. *intrafusal*

b. Stimulation of gamma efferent fibers causes contraction of *polar / equatorial* regions of muscle fibers.

b. *polar*

c. The spindle as a whole will be *shortened / stretched*.

c. *shortened*

d. If the extrafusal fibers do not contract at the same time, the receptors in the noncontractile equatorial region of the spindle will be *shortened / stretched*.

d. *stretched*

e. Excitation of the primary and secondary receptors stimulates the muscle's lower _____.

e. *motoneurons*

f. The muscle and the spindle are thus *shortened / stretched*.

f. *shortened*

g. What then happens to the receptors?

g. *no longer stimulated*

To the extrafusal fibers?

cease to contract

To the muscle spindle and muscle complex?

in a state of equilibrium

Characteristics	Muscles:	Smooth	Skeletal
more likely to be activated by chemical stimulation		✔	
rather independent of nerve supply		✔	
dependent on nerve supply			✔
more subject to the stretch reflex			✔

Muscle Characteristics	Loss of Nerve Supply:	No Loss	Peripheral	Cerebral	Afferent
stretch reflex can be elicited		✔		✔	
stretch reflex cannot be elicited			✔		✔
muscle tone present		✔		✔	
muscle tone absent			✔		✔

The following steps are part of a stretch reflex.
Number them in correct sequence (1-6)

The muscle contracts.   _____

The muscle is stretched.   _____

Impulses travel out of spinal cord to the muscle.   _____

Impulses reach spinal cord via afferent fibers.   _____

Muscle spindle initiates a train of impulses.   _____

Afferent fibers synapse with lower motoneurons.   _____

Description	Reflex Arc:	2-Neuron	3-Neuron
withdrawing hand from a painful stimulus			✔
stretch reflex, knee jerk reflex		✔	
involves sensory and motor neurons		✔	✔
involves internuncial neurons			✔
information may reach cerebellum and/or cerebrum			✔

Notes: The familiar "knee jerk" elicited by the physician's hammer is an example of a stretch reflex. The tap of the hammer slightly lengthens the extensor muscle, thereby stimulating its stretch receptors. They cause the muscle to contract, and your leg kicks (without your help).

A *spasm* is a sudden involuntary contraction of a muscle or group of muscles.
Clonic spasm. Alternating rigidity and relaxation of muscles.
Tonic spasm. Prolonged rigidity of muscles.

Quote: "Extensors are principally antigravity muscles, straightening our back and lower limb joints, among others, to keep us upright in the gravitational field in which we live. A sudden unexpected lengthening of an extensor muscle is interpreted as a possible loss of antigravity tone and sets off an immediate shortening reaction (stretch reflex) to 'take up the slack.'"

Diamond, Scheibel, and Elson, 1985

Description	Pathway for:	Pain Temperature	Pressure Crude Touch
receptors in the skin		✔	✔
afferent neuron cell bodies in dorsal ganglia		✔	✔
bifurcates in spinal cord			✔
ascends as ventral spinothalamic tract			✔
ascends as lateral spinothalamic tract		✔	
synapses with third-order neurons in thalamus		✔	✔
reaches postcentral gyrus of cerebral cortex		✔	✔

Description	Fasciculus:	Cuneatus	Gracilus
ascending tract		✔	✔
receives fibers from lower body			✔
receives fibers from upper body		✔	
medial part of dorsal column			✔
lateral part of dorsal column		✔	
pathway for proprioception		✔	✔
pathway for vibration		✔	✔
pathway for fine touch		✔	✔
terminates at synapse with 2nd order neurons in medulla		✔	✔
fibers decussate and give rise to medial lemniscus which ascends to thalamus		✔	✔

1. The sense of awareness of body parts and their movement and position is called _____.

   1. _____

2. Two-point discrimination and stereognosis are possible because of the sense of fine _____.

   2. _____

No. 5-36 THE PYRAMIDAL PATHWAY
Text pages 404–405

1. The pyramidal pathway is also called the *voluntary /*
   involuntary motor or _____ pathway.

   1. _____
      _____

2. The pyramidal pathway primarily supplies voluntary
   muscles of the head, neck and *trunk / limbs*. It also
   supplies nuclei of the _____ nerves.

   2. _____
      _____

3. The giant pyramidal cells of Betz, from which most fibers
   of the pyramidal pathway arise, are in the _____
   cortex.

   3. _____

4. Most fibers of the corticospinal tract, after converging
   to form pyramids in the medulla oblongata, decussate
   to descend in the lateral funiculus as the _____
   pyramidal tract.

   4. _____

5. The remaining fibers of the corticospinal tract, those
   which do not decussate, descend in the ventral funiculus
   as the _____ pyramidal tract.

   5. _____

Note: Because the pyramidal system is called a direct motor pathway, the division of the corticospinal tract
into direct and crossed pyramidal tracts may seem incongruous. Physiologically the pyramidal tract is
a direct motor pathway because the axons from the cerebral cortex extend directly, without any
additional synapses, to the area of the motor neurons. *Anatomically* the tracts are *direct* or *crossed*, but
physiologically they are both *direct*.

Structures Primary Cortical Representation:	Unilateral	Bilateral
head, neck, and trunk		✔
arms and legs	✔	
most musculature of pharynx, larynx, and soft palate		✔
facial regions which are bearded in adult male	✔	
facial regions which are unbearded in adult male		✔

6. Fibers of the corticobulbar tract terminate *at / above /*
   below the level of the pyramids.

   6. _____

7. The pyramidal system has some control over all *voluntary /*
   involuntary / voluntary and involuntary musculature.

   7. _____

8. The pyramidal system is particularly important in the
   mediation of *fine / gross* motor movement.

   8. _____

9. If excitatory fibers of the pyramidal system were
   damaged, what would be the effects on
   complex muscle movement?

   9. _____
      _____

10. If inhibitory fibers of the pyramidal system were
    damaged, the results would be hyperreflexion and
    _____.

    10. _____

No. 5-36 cont'd

Quote: "Upper motor neuron paralysis occurs when there is damage to the corticospinal tract anywhere along its path: the cell bodies in the precentral gyrus or their descending axons in the internal capsule, brainstem, or spinal cord. The most common site of injury is in the cerebral hemisphere, before the decussation. Injury results most often when an artery becomes stopped up and the neurons, deprived of their oxygen supply, die, producing what's known as a cerebrovascular accident (CVA) or, in popular language, a stroke. If the site affected is above the motor decussation, then the signs and symptoms will be seen in the muscles on the opposite side of the body. If the injury is after the decussation, say a cut in the left half of the spinal cord, then the ensuing paralysis will be on the same side of the damage."

M. Liebman, 1979

No. 5-37 THE EXTRAPYRAMIDAL PATHWAY
Text pages 405–406

1. The extrapyramidal system includes all the descending pathways except the _____ tracts.

   1. _____

2. The extrapyramidal tract, sometimes called the alternate pathway for motor impulses, is primarily a *coordinating / generating* pathway.

   2. _____

3. Whereas the pyramidal pathways from the motor cortex to the spinal cord are physiologically *direct / indirect*, the extrapyramidal pathways are physiologically *direct / indirect*.

   3. _____
   _____

4. The reticular substance of the brain stem may *excite / inhibit / excite or inhibit* the activity of motor neurons.

   4. _____

Description/Function Tracts:	Vestibulospinal	Rubrospinal	Tectospinal	Olivospinal
impulses originate in cerebral cortex	✔	✔	✔	✔
mediates visual and auditory reflexes			✔	
integrates and coordinates voluntary motor activity				✔
coordinates reflexive postural behavior		✔		
maintains equilibrium and posture	✔			
begins in vestibular nucleus	✔			
begins in colliculi (quadrate bodies)			✔	
begins in olivary nucleus				✔
begins in red nucleus		✔		

Quote: "The voluntary movements initiated by the extrapyramidal system have been described as chiefly *axial*, such as bowing, walking, rolling, kneeling, sitting and standing. Much of our understanding of extrapyramidal voluntary movements has come from the study of apraxias, disorders in the execution of certain types of movements on command that can be explained by damage to fibers linking the language areas in the left cerebral hemisphere because of injury to the corpus callosum, with the disruption of tracts connecting the left and right cerebral hemispheres. Such patients, upon command to 'show me how you comb your hair,' can usually carry out the command with the right arm only because the command cannot reach the motor areas in the right hemisphere which initiate voluntary movements via the pyramidal system on the left side of the body. However, commands to walk backwards, kneel, or bend the head down can be executed because the extrapyramidal system on one side of the brain controls axial muscles of the neck and body on both sides."

Jacob, Francone, and Lossow, 1978

No. 5-38 THE NERVOUS CONTROL OF RESPIRATION
Text pages 406–408

1. The regulation of the depth and rate of breathing may
   be *voluntary / involuntary / both.*

2. Alveolar ventilation, which maintains relatively constant
   concentrations of oxygen and carbon dioxide in the blood,
   is regulated by the respiratory center located in the
   _____.

3. What are the three principal stimuli
   responsible for respiratory regulation?

4. The maintenance of normal respiratory rate and depth
   which seems to be regulated by the stimulation of sensory
   endings in the lungs is called the _____ reflex.
   Maintaining the rate of respiration is also a function of
   the _____ center.

5. Chemoreceptors respond directly to the chemistry of the
   blood / inhailed air / both.

6. When stretch receptors in the lung tissue, visceral pleura,
   and the bronchial tree are relaxed the frequency of impulses
   decreases, thus signaling the onset of *inspiration / expiration.*

7. When these stretch receptors are stimulated the frequency of
   impulses increases, thus signaling the inhibition of *inspiratory /
   expiratory* forces and signaling the onset of passive *expiratory /
   inspiratory* forces.

1. _____

2. _____

3. _____
   _____
   _____

4. _____
   _____

5. _____

6. _____

7. _____
   _____

No. 5-39 THE NERVOUS CONTROL OF THE TONGUE, PHARYNX, LARYNX, SOFT PALATE, AND MUSCLES OF MASTICATION
Text pages 408-410

The Tongue

1. The intrinsic and extrinsic muscles of each half of the
   tongue are supplied by the _____ (C XII) nerve,
   whose nucleus of origin has *unilateral / bilateral* cortical
   representation.

2. The palatoglossus, which receives its motor fibers from
   the _____ (C XI) nerve by way of the pharyngeal
   plexus, may be considered a muscle of the *palate / pharynx.*

3. What appears to be the role of the
   muscle spindles in the tongue?

4. What are the three sensory nerves of the tongue?

1. _____
   _____

2. _____
   _____

3. _____
   _____

4. _____
   _____
   _____

The Muscles of Mastication

1. The mandibular elevators are supplied by the _____ (C V) nerve.

2. The mandibular depressors are supplied by the _____ (C V) and _____ (C XII) nerves.

3. The cortical representation of the nuclei of origin of these nerves is *unilateral / bilateral*.

1. _____

2. _____

3. _____

The Pharynx

1. The pharynx is innervated by the motor branch of the pharyngeal plexus, which begins as two rami of the _____ (C X) nerve.

2. Sensory innervation of the pharynx is supplied by the _____ (C IX) nerve.

1. _____

2. _____

The Soft Palate

1. The motor fibers for the muscles of the soft palate are derived from the *mandibular / maxillary* branch of the trigeminal (C V) nerve.

2. The motor fibers to the uvula are supplied by the _____ (C XI) nerve.

3. The sphenopalatine ganglion, which supplies the palate with both sensory and motor fibers, receives sensory fibers from the *mandibular / maxillary* branch of the trigeminal (C V) nerve and from the _____ (C VII) nerve.

1. _____

2. _____

3. _____
_____

The Larynx

1. Most of the intrinsic muscles of the larynx are supplied by the recurrent branches of the _____ (C X) nerve.

2. The extrinsic muscles of the larynx are innervated by the _____ (C V), _____ (C VII), and _____ (C XII) nerves.

1. _____

2. _____
_____
_____

adrenal glands   hormones   pituitary gland
endocrine system   pancreas   thyroid gland
gonads   parathyroid glands   thyroxin (thyroxine)

1. comprised of glands secreting substances into the fluids of the circulatory system

1. _____

2. endocrine gland secretions that regulate specific functions of other organs or tissues of the body

2. _____

3. located behind the stomach

3. _____

4. located on the upper surface of the kidneys

4. _____

5. two lobes, one on either side of the larynx; connected by an isthmus

5. _____

6. located in the sella turcica at the base of the brain

6. _____

7. two pairs of pea-sized glands behind the thyroid gland

7. _____

8. a hormone secreted by the thyroid gland

8. _____

9. testes and ovaries

9. _____

10. site of Islets of Langerhans which produce insulin

10. _____

11. also called hypophysis (below + to grow)

11. _____

12. a hormone containing iodine; increases metabolic rate

12. _____

13. regulates the function of other endocrine glands and the growth and development of the body

13. _____

14. regulate levels of calcium and phosphorus in the blood

14. _____

15. secrete adrenalin (adrenin, epinephrine) which stimulates sympathetic nervous system and regulates carbohydrate metabolism and blood sugar level

15. _____

16. responsible for physical changes occurring during puberty

16. _____

17. produces enzymes that aid digestion

17. _____

18. deficiencies of this hormone may produce cretinism and myxedema

18. _____

19. overactivity of cortex may result in exaggeration of male secondary sex characteristics

19. _____

Chapter 6
Hearing

No. 6-1 THE NATURE OF SOUND: SIMPLE HARMONIC VIBRATION
Text pages 415–417

1.

Characteristics of Vibration	Reed:	Large	Small
a.	greater amplitude		
b.	higher frequency		
c.	will produce shorter sound waves		
d.	longer period		

2. What kind of motion is exemplified by vibrating reeds?

2. _____

3. What kind of wave will be produced by a reed vibrating in a vacuum?

3. _____

4. If a reed is vibrating at 60 Hz, what is its period?

4. _____

5. Undisturbed air is in a state of _____. Although the air particles are moving randomly their density remains relatively _____.

5. _____
_____

6. Air tends to flow from regions of higher pressure to regions of lower pressure because it is a/an _____.

6. _____

7. Air exhibits inertial properties because it has _____.

7. _____

8. As a reed is set into vibration, movement to the right will cause the air molecules on the right to be _____. As it moves back to the left those molecules will then become _____.

8. _____
_____

9. The spherical waves produced by the vibrating reeds will be *transverse / longitudinal.*

9. _____

10. When the reeds vibrate, the disturbed molecules of air *move in a wavelike fashion / impart energy to succeeding molecules.*

10. _____

11. If you were to initiate vibration of a reed, the restoring force and acceleration would be proportional to the _____.

11. _____

No. 6-2 SINE WAVES
Text pages 415–417

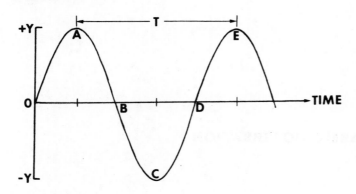

Point(s) Representing:	A B C D E
1. maximum displacement	A B C D E
2. compression	A B C D E
3. rarefaction	A B C D E
4. equilibrium	A B C D E

*On the illustration, indicate two points which represent 360°
of phase change and label with λ, the symbol for wavelength.*

5. What is the amplitude of the wave?   5. _____

6. If T = .1 second, what is the frequency?   6. _____

7. If T = .002 second, what is the frequency?   7. _____

8. If the frequency is 2000 Hz, what is T?   8. _____

9. If the frequency is 10,000 Hz, what is T?   9. _____

Note: Sine is derived from the Latin word *sinus* meaning gulf or bosom (a translation of an Arabic word referring to the bosom of a garment).

No. 6-3 THE NATURE OF SOUND (VOCABULARY)
Text pages 415–420

amplitude	free vibration	phase
Brownian movement	frequency	rarefaction
cancellation	Hooke's Law	resonance
compression	inertia	root-mean-square amplitude
critical damping	maintained vibration	simple harmonic motion
damping	mass	sinusoidal motion
displacement	period	vibratory motion
forced vibration	periodic	wave motion

1. the distance between the position of equilibrium and the position of the body at a specified instant   1. _____

2. a body's absorption and emission of energy at the same frequency band   2. _____

3. a quantity of matter; the physical measure of the principal inertial properties of a body   3. _____

4. the time elapsed during a single complete vibration   4. _____

5. induced oscillation of a body at an unnatural frequency   5. _____

6. causing a decrease in amplitude of successive waves or oscillations   6. _____

7. movement to and fro along a path in such a manner that there is a restoring force which increases with displacement and is always directed toward the position of rest

7. _____

8. an increase in the density of molecules

8. _____

9. the property of all matter by which it resists any change in its state of motion

9. _____

10. projected uniform circular motion (2)

10. _____
_____

11. random movement of air particles

11. _____

12. periodic vibration of a body at its natural frequency which once initiated continues until the energy has been dissipated

12. _____

13. maximum displacement of the body from its position of equilibrium; usually equal to one-half the total extent of vibratory motion

13. _____

14. restoring force is proportional to displacement; strain is proportional to stress

14. _____

15. constant induced oscillation of a body at an integral multiple of its natural frequency, thus permitting sustained, constant amplitude vibration of the body

15. _____

16. reduction of amplitude of a sound (ideally to zero) when two sine waves having the same amplitude and frequency are 180° out of phase

16. _____

17. a decrease in the density of molecules

17. _____

18. displaced body returns to its position of equilibrium without going beyond it

18. _____

19. number of complete vibrations or cycles per unit time; usually measured in vibrations or cycles per second (Hertz, Hz)

19. _____

20. portion of a cycle through which a vibrating body has passed up to a given instant; usually expressed in terms of degrees of a circle

20. _____

21. generated by a vibrating body when it is coupled to or immersed in an elastic medium

21. _____

22. gives the average amplitude for the entire period of a sinusoid

22. _____

23. movement of a vibrating element

23. _____

24. occurring in equal time intervals

24. _____

No. 6-4 SOUND WAVES
Text page 418

Description/Examples Waves:	Transverse	Longitudinal
transmitted by medium having compressional elasticity		✔
transmitted by a medium that will support shearing stress	✔	
transmitted by all forms of matter (solids, liquids, gases)		✔
sound waves in the air		✔
waves on a string	✔	
displacement of individual particles is perpendicular to direction of propagation	✔	

1. In a progressive longitudinal wave

    a. the direction of particle movement is *parallel / perpendicular* to the direction of propagation of the disturbance

    b. each molecule executes the same motion as the preceding molecule *simultaneously / a bit later*.

1.

    a. _____

    b. _____

No. 6-5 PHASE RELATIONSHIPS OF SOUND WAVES
Text pages 418–419

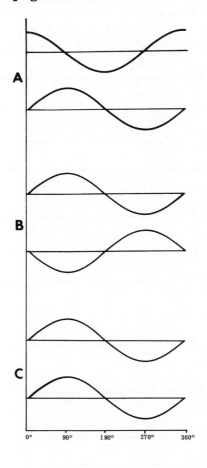

	Description Waves:	A	B	C
1.	waves are in phase	A	B	C
2.	waves are 180° out of phase	A	B	C
3.	waves are 90° out of phase	A	B	C
4.	will result in cancellation	A	B	C
5.	will result in maximum reinforcement	A	B	C
6.	will result in partial reinforcement	A	B	C
7.	when one wave is in a state of equilibrium, the other is in a state of maximum compression or rarefaction	A	B	C
8.	regions of compression and rarefaction are congruent	A	B	C
9.	one wave is maximally compressed while the other is maximally rarefied	A	B	C
10.	sum of the amplitudes will always be zero	A	B	C
11.	sum of the amplitudes is double the amplitude of one wave	A	B	C

Note: The phase difference between two waves (x and y): 0° 90° 180°

Resultant amplitude: $x + y$ $\sqrt{x^2 + y^2}$ $x - y$

1. What is the symbol for wavelength?

2. A complete cycle of simple harmonic motion represents a phase change of *180° / 360°*.

3. At room temperature, the velocity with which sound waves travel through the air is *130 / 1,130 / 11,300* feet per second.

4. If a tuning fork is vibrating at 256 Hz, the length of the sound wave is approximately _____ feet.

5. If the length of a sound wave is two feet, what is its frequency?

6. If the amplitude of sound wave A is two feet and the amplitude of sound wave B is four feet, the energy of sound wave B is *two / four / eight* times that of wave A.

7. A sound wave ten feet from the source has an intensity of I.

    a. A sound wave of 1/4 *I* will be _____ feet from the source.

    b. A sound wave of 1/16 *I* will be _____ feet from the source.

    c. A sound wave of 1/25 *I* will be _____ feet from the source.

8. The rate of energy flow per unit of area of surface receiving the flow is called _____.

1. _____

2. _____

3. _____

4. $\lambda = v/f$ _____

5. $f = v/\lambda$ _____

6. _____

7. $I = l/d^2$

    a. _____

    b. _____

    c. _____

8. _____

FIGURE 6.1 GRAPHS OF SIMPLE HARMONIC VIBRATION SHOWING THAT A PHASE CHANGE OF 360° REPRESENTS ONE COMPLETE CYCLE OF VIBRATION, REGARDLESS OF THE PERIOD.

FIGURE 6.2 SCHEMATIC ILLUSTRATION OF THE PRINCIPLE OF THE INVERSE SQUARE LAW.

A = area   r = radius

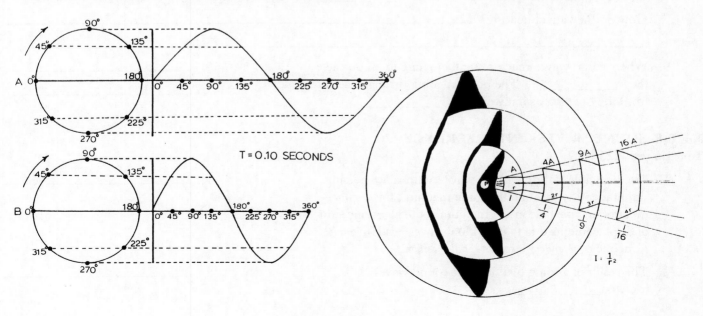

No. 6-7 SOUND WAVES: REFLECTION; DIFFRACTION
Text pages 422–425

1. When sound encounters a boundary between two propagating mediums having different physical properties it is _____. Such a boundary is called a/an _____.

   In room acoustics these boundaries are the _____, _____, and _____.

   1. _____

   _____

   _____

   _____

   _____

2. A compression wave is reflected as a compression at the *barrier / free end.*

   2. _____

3. According to the law of reflection, the angle of incidence is *smaller than / equal to / greater than* the angle of reflection.

   3. _____

4. When incident spherical waves strike a wall

   a. they are reflected back towards the _____.

   b. the reflected waves are *spherical / not spherical.*

   c. the reflected waves appear to be coming from a point behind the _____. This point is called the _____ of the source.

   4.

   a. _____

   b. _____

   c. _____

   _____

5. The persistence of sound resulting from multiple reflections within an enclosed space is called _____.

   5. _____

6. The principle stating—every point of an advancing wave front is the center of a fresh disturbance and a source of new smaller waves is Huygen's principle of _____ wavelets. *See text Fig. 6-16.*

   6. _____

7. A change in the direction of propagation of sound waves due to the presence of obstacles in their path is called _____.

   7. _____

8. Regions lacking in wave energy due to the presence of obstacles in the wave pathways are called _____.

   How do they affect hearing? _____

   _____

   8. _____

9. When sound waves reach a small opening in an obstacle they _____. The portion going through radiates from the opening as if it were the _____.

   9. _____

   _____

No. 6-8 SOUND WAVES: INTERFERENCE
Text pages 425–427

1. The principle stating—if two wave motions are passing simultaneously through the same medium, the displacement of any particle in the medium is the algebraic sum of displacements due to individual waves—is Huygen's principle of *secondary wavelets / superposition.*

   1. _____

2. The result of adding two or more sound waves is called _____.

   2. _____

No. 6-8 cont'd

Characteristics/Examples Interference:	Constructive	Destructive
cancellation		✔
reinforcement	✔	
the sum of displacements produced by two or more individual waves	✔	✔
two sound waves having same frequency and phase, but different sources, coincide	✔	✔
two compressions coincide	✔	
two rarefactions coincide	✔	
a compression and a rarefaction coincide		✔
created when a sounding tuning fork is rotated on an axis through its stem	✔	✔
fixed patterns generate standing waves	✔	✔
heard as an increase in loudness	✔	
heard as a decrease in loudness		✔
may be perceived as beats when two waves, with only slight frequency differences, go in and out of phase	✔	✔

Note: The principle of superposition explains why we as listeners can be in the presence of a number of people all talking at the same time (as is usually the case) and yet we can listen to just one speaker or to the entire cacophony.

FIGURE 6.3 TRANSVERSE STANDING (STATIONARY) WAVES.

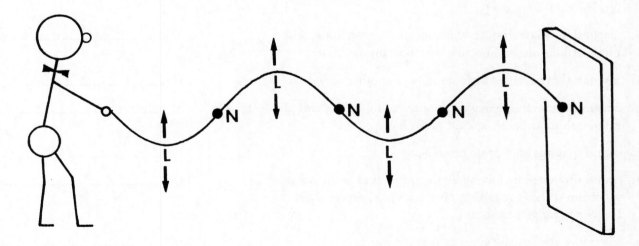

With proper tuning, waves generated on a rope will appear to be standing still. Regions where no movement occurs are called *nodes*, and regions where maximum movement takes place are called *loops*. Similarly, standing waves can be generated in fluid media contained by reflecting surfaces.

amplitude spectrum	intensity	reflected wave
average power	inverse square law	standing wave
beats	law of reflection	stationary wave
complex sound	masking	steady-state sound
elasticity	noise	transient sound
elastic medium	octave	wave front
frequency	overtone	waveform
fundamental frequency	partial	wavelength
harmonic	pure tone	white noise
incident wave	ray of sound	

1. the specific direction in which a sound wave travels

2. the advancing disturbance generated by a sound wave

3. the intensity of a sound wave varies inversely as the square of the distance from the source; intensity =
$$\frac{1}{(\text{distance from the source})^2}$$

4. a sound composed of more than one frequency

5. the greatest common divisor of the component frequencies of a periodic wave or quantity

6. any physical component of a complex sound

7. a physical component of a complex sound having a frequency higher than the fundamental frequency

8. frequency range × intensity

9. a partial whose frequency is an integral multiple of the fundamental frequency

10. a sound in which frequency composition, amplitude, and phase relationships of partials are constant over time

11. a sound that obscures the audibility of another sound

12. a sound whose frequency composition, amplitude, and phase relationships are not constant over time

13. sound that has little or no periodicity

14. sound produced by mixing all the pure tones in the audible spectrum without regard to phase; average power over frequency range is constant

15. a sound wave initially generated at the source

16. a sound wave that has been cast or thrown back

1. _____

2. _____

3. _____

4. _____

5. _____

6. _____

7. _____

8. _____

9. _____

10. _____

11. _____

12. _____

13. _____

14. _____

15. _____

16. _____

No. 6-9 cont'd

17. the wave resulting when two sound pressure waves having equal amplitude and frequency, travel the same path but in opposite directions

17. _____
_____

18. $= \dfrac{\text{the velocity in feet per second}}{\text{the frequency in cycles per second (Hz)}}$

18. _____

19. $= \dfrac{\text{the velocity in feet per second}}{\text{wavelength in feet}}$

19. _____

20. the capacity of matter to recover from distortion when external force is removed

20. _____

21. matter (gases, liquids, solids) which has restoring forces and thus the ability to conduct sound

21. _____

22. a graphic representation of amplitude as a function of frequency

22. _____

23. when audible, perceived as very regular increases and decreases in loudness; periodic variations resulting from the superposition of waves having different frequencies

23. _____

24. commonly used term for sinusoidal note

24. _____

25. rate of energy flow per unit of area of surface receiving the flow

25. _____

26. any interval of two frequencies having a ratio of 2:1

26. _____

27. any undesirable sound

27. _____

28. sound with an instantaneous amplitude that varies over time in a random manner

28. _____

29. graphic representation of displacement or amplitude as a function of time

29. _____

30. angle of incidence = angle of reflection

30. _____

Components of a Complex Tone	125 Hz	250 Hz	375 Hz	500 Hz
partials	✔	✔	✔	✔
overtones		✔	✔	✔
harmonics	✔	✔	✔	✔
fundamental frequency	✔			
first harmonic	✔			
second overtone			✔	
third harmonic			✔	

Vibration Patterns on a Stretched String	Harmonics	Overtones
2 nodes; 1 loop (fundamental frequency)	0 (1) 2 3 4	(0) 1 2 3 4
3 nodes; 2 loops	0 1 (2) 3 4	0 (1) 2 3 4
4 nodes; 3 loops	0 1 2 (3) 4	0 1 (2) 3 4
5 nodes; 4 loops	0 1 2 3 (4)	0 1 2 (3) 4

Characteristics/Examples Sounds:	Steady-State	Transient
complex tones, constant over time	✔	
a change in steady-state		✔
usually a consonant		✔
a prolonged vowel sound	✔	
graphically represented by waveform	✔	✔
a sustained musical tone	✔	
the general nature of connected speech		✔

Description Noise:	Gaussian	White	Pink
has all frequencies within specified range without regard to phase; average power is constant		✔	
with each doubling of frequency amplitude decreases by one-half			✔
has an instantaneous amplitude that varies in the manner of a normal distribution	✔		
frequently used as masking noise		✔	
frequently used in studies of audition			✔

1. Noise may be generated when air passing through a constriction creates *turbulence / reverberation*.

1. _____

2. Noise may be generated when a body vibrates *periodically / aperiodically*.

2. _____

No. 6-11 RESONANCE AND FILTERS
Text pages 431–433

1. An object resonating at its natural frequency is acting as an acoustic *amplifier / filter*.

    1. _____

2. The process of changing the natural frequency of a resonator is called _____.

    2. _____

3. If, by adding water to a bottle, you shorten the air column, the frequency of the edge tone will be *increased / decreased*.

    3. _____

Description/Characteristics -pass Filter:	Low-	High-	Band-
usually attenuates high frequencies	✔		
usually attenuates low frequencies		✔	
passive filter	✔	✔	✔
attenuates frequencies on either side of a specific frequency range			✔

No. 6-12 AMPLIFIERS
Text pages 433–434

1. An amplifier is a device that uses a small amount of power to control a large amount of _____.

    1. _____

Description/Characteristics	Nonlinearity	Limited Frequency Range
may cause distortion	✔	✔
20–20,000 Hz		
output not representative of input	✔	✔
500–8000 Hz		✔
may add higher harmonics of input	✔	
common problem of amplifiers	✔	✔

Note: The *voltage gain* of an amplifier is the ratio of output voltage to input voltage. If the input voltage is 1 millivolt (.001 volts) and the output voltage is 1 volt, the voltage gain is 1000 (1/.001) or 60 dB ($dB = 20 \, Log \, V_2/V_1$). Voltage gain, the increase of voltage occurring in the amplifier, is not an indication of the amount of work an amplifier is capable of.

The *current gain* of an amplifier is the ratio of output current to input current (measured in amperes or milliamperes).

Electrical power (in watts or milliwatts) is the product of *current* times *voltage*. Thus, the *power gain* of an amplifier is the *current gain* times *voltage gain*. It specifies in watts or milliwatts how much work an amplifier is capable of.

Depending on their design, amplifiers may be classified as power amplifiers or voltage amplifiers. An amplifier that drives a loudspeaker is a *power amplifier*, while an amplifier used with a CD player or a microphone is a *voltage amplifier*. Audio amplifiers usually utilize a *preamplifier* (voltage amplifier) followed by a power amplifier.

1 = 1	3 = 3	3 $= 3^1 =$ 3	10 $= 10^1 =$ 10
1 + 1 = 2	3 + 3 = 6	3 × 3 $= 3^2 =$ 9	10 × 10 $= 10^2 =$ 100
2 + 1 = 3	6 + 3 = 9	3 × 3 × 3 $= 3^3 =$ 27	10 × 10 × 10 $= 10^3 =$ 1000
(a)	(b)	(c)	(d)

1. (a) and (b) are examples of _____ scales.

2. (c) and (d) are examples of _____ scales.

3. The base of (c) is _____.

4. The base of (d) is _____.

5. The ratio of the successive products in (c) is ____:____.

6. The ratio of the successive products in (d) is ____:____.

7. The intervals between successive values are equal and linear in *(a) and (b) / (c) and (d)*.

8. Any scale whose successive units are multiplied by a specific base is called an exponential or _____ scale.

9. The exponents in (c) and (d) form a/an *interval / ratio* scale.

10. In (c), 27 × 9 would equal 3 × 3 × 3 × 3 × 3, or 3 to what power?

    You can quickly achieve this answer by adding the _____.

11. In (d), 1,000 ÷ 100 would equal 10 × 10 × 10 ÷ 10 ÷ 10, or 10 to what power?

    You can quickly achieve this answer by subtracting the _____.

12. The intensity ratio of the loudest tolerable sound to the just audible sound is 100,000,000,000,000 to 1 or 10 to what power?

13. The ratio of 100,000,000,000,000 to 1 may thereby be expressed as _____: 1 or _____ bels.

14. The decibel is one-_____ the power ratio of a bel.

15. Why is the decibel a more useful measure than the bel?

16. decibels = 10 × logarithm (to base 10) × $\dfrac{\text{intensity no. 2}}{\text{intensity no. 1}}$

    This formula expresses a/an _____ or _____ ratio.

17. Intensity is a measure of energy flow per unit of _____ per unit of _____.

18. Energy per second, measured in watts per square centimeter is called _____.

19. $\dfrac{10^{-16}\text{ watts}}{\text{cm}^2}$ is a standard reference for _____.

1. _____

2. _____

3. _____

4. _____

5. _____ : _____

6. _____ : _____

7. ____( ) and ( )____

8. _____

9. _____

10. _____

    _____

11. _____

    _____

12. _____

13. _____

    _____

14. _____

15. _____

16. _____

    _____

17. _____

    _____

18. _____

19. _____

20. A negative exponent is the reciprocal of the same number with a positive exponent. $1/10 \times 1/10 \times 1/10 \times 1/10$ may be expressed as 10 to what power?

20. _____

21. The power of a sound is proportional to the square of the _____.

21. _____

22. decibels = $20 \times$ logarithm (to the base 10) $\times \dfrac{\text{pressure no. 2}}{\text{pressure no. 1}}$

The above formula expresses _____ ratios in decibels.

22. _____

23. Whereas force may be defined as a push or pull, pressure is defined as force per unit _____.

23. _____

24. If the force remains the same and the unit area decreases, pressure will *increase / decrease*.

24. _____

25. If you were teaching this class, how would you demonstrate the difference between force and pressure?

25. _____
_____
_____

26. $\dfrac{.0002 \text{ dynes}}{\text{cm}^2}$ is the standard reference for _____.

A dyne is a centimeter-gram-second unit of force.

26. _____

27. The differences in dB between any pressure and .0002 dynes/cm^2 is known as _____, abbreviated _____.

27. _____
_____

No. 6-14 THE EAR: INTRODUCTION
Text pages 435–437

1. What is the approximate range of audibility of the human ear?

1. _____ Hz to _____ Hz

2. What are the three anatomical divisions of the ear?

2. _____
_____
_____

3. What are the two functional divisions of the ear? What are the roles of each division?

3. _____
_____
_____
_____

1. What is the primary function of the external   1. _____
   auditory meatus (ear canal)?

2. The junction of the cartilaginous and bony framework of the   2. _____
   ear canal is called the *orifice / isthmus*.

3. The tympanic membrane marks the termination of the   3. _____
   cartilaginous / bony portion of the ear canal.

4. There is greater variability of the diameter of the *lateral /*   4. _____
   medial portion of the ear canal.

5. Aside from the fact that small children   5. _____
   often stuff strange things in their ears,
   why are foreign materials more likely   _____
   to collect in the ears of children than
   in the ears of adults?   _____

6. What is the proper term for "earwax?"   6. _____

   What are its functions?   _____

   What type of glands produce it?   _____

7. The cilia of the external auditory meatus are located   7. _____
   laterally / medially.

8. The angular direction of the sound source in relationship   8. _____
   to the listener is called the *azimuth / auricle*.

9. If a sound is generated in a free field there should be no   9. _____
   incident / reflected waves.

10. Binaural hearing contributes to the *amplification / localization*   10. _____
    of sounds.

11. Because the external auditory meatus is a tube closed at   11. _____
    one end, it should resonate at a frequency whose wave-
    length is four times the *length / diameter* of the tube.

12. The resonant peak of the meatus is near *1000 / 2000 /*   12. _____
    4000 Hz.

13. Why does the meatus have a wider   13. _____
    range of resonance than an
    ordinary tube closed at one end?   _____

    _____

14. Is the resonant peak of the concha higher or lower than that   14. _____
    of the meatus?

15. At frequencies of 2000 to 5000 Hz, resonance and head   15. _____
    effects increase the sound pressure level at the drum
    membrane by approximately *3 / 9 / 15* decibels.

No. 6-15 cont'd

Note: Transduction, meaning to lead across, is the translation of one form of energy to another.

oto- ot-	combining forms relating to the ear
otitis media	inflammation (not necessarily infection) of middle ear
otalgia	earache
otoscope	instrument for examination of the ear
otosclerosis	a hardening of a portion of the ear
otorrhea	a discharge (through a perforated eardrum) from the ear
ototoxic	poisonous to the ear, e.g., drugs which may cause deafness
Medical Specialties:	otology, otolaryngology, otorhinolaryngology

No. 6-16 THE EFFECTS OF OBJECTS IN A SOUND FIELD
Text pages 438–439

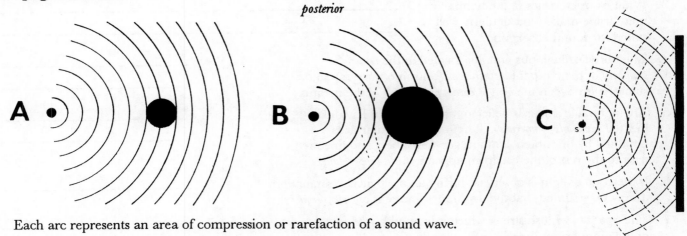

Each arc represents an area of compression or rarefaction of a sound wave.

Description	Illustration:	A	B	C
waves reflected by an object much larger than the wavelength		A	B	(C)
waves passing by an object approximately the same size as wavelength		(A)	B	C
creation of a "sound shadow" when sound wave encounters an object somewhat larger than the wavelength		A	(B)	C

1. Imagine that the object in B is a head as seen from above. Note *posterior* at the top of the illustration.

   a. The sound pressure level will be greater at the *right / left* eardrum.

   1. a. _____

   b. Would the intensity difference be greater at 1000 Hz or 8000 Hz?

   b. _____

   Why? _____

1. The tympanic membrane (eardrum) forms the medial boundary   1. _____
    of the _____ and the lateral boundary of the _____.   _____

2. Which of the above best represents the usual angle of the   2. _____
    tympanic membrane in an adult?   _____
    In a newborn infant?

    What accounts for the difference   _____
    between adults and infants?   _____

3. What characteristics of the tympanic   3. _____
    membrane make it particularly well   _____
    suited for sound absorption?

4. The fibrocartilaginous annulus located at the *center /*   4. _____
    periphery of the tympanic membrane, fits into the tympanic   _____
    sulcus, a groove in the *bony / cartilaginous* wall of the meatus.

5. The tympanic sulcus is deficient at the notch of Rivinus,   5. _____
    located *superiorly / inferiorly*. The small triangular area   _____
    bounded by this notch contains very few fibers and in con-
    trast to the rest of the eardrum appears quite *limp / taut.*

6. If the cone of light is observed during an otological examination   6. _____
    it may be assumed that the drum membrane is *concave / convex.*

7. The opaque, whitish streak which is formed by the attachment   7. _____
    of the manubrium (handle) of the malleus to the drum membrane,
    is called the *cone of light / malleolar stria.*

8. The cone of light radiates from a concavity in the center of   8. _____
    the drum membrane called the "navel" or *umbo / incus.*

9. Light is also reflected from the malleolar prominence that is 9. _____
    formed by the attachment of the lateral process of the
    malleus / stapes to the drum membrane.

10. When the tympanic membrane is divided into quadrants, the   10. _____
    vertical axis goes through the _____ and is bisected   _____
    by the horizontal axis at the level of the _____.

Description	Pars Tensa	Pars Flaccida
most of the tympanic membrane	✔	
small triangular area bounded by notch of Rivinus		✔
below the malleolar folds	✔	
once called Schrapnell's membrane		✔
may contribute to the equalization of pressure between the external and middle ear		✔

FIGURE 6.4 TYMPANIC MEMBRANE AND OSSICULAR CHAIN.

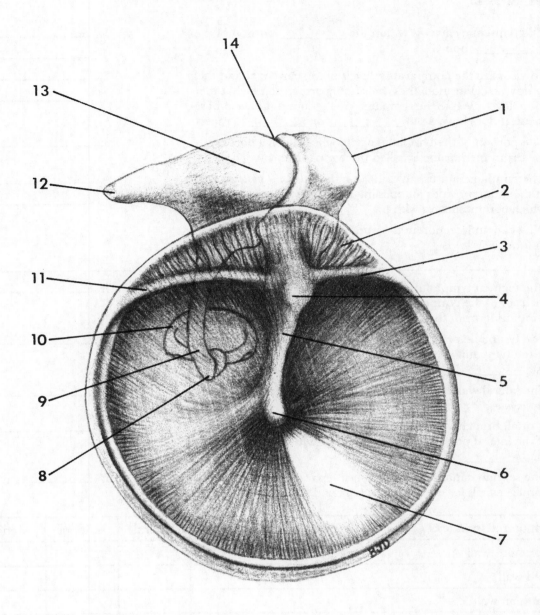

Identify:

1. head of malleus
2. pars flaccida
3. anterior malleolar fold
4. lateral malleolar process
5. manubrium of malleus
6. umbo
7. cone of light
8. lenticular process of incus
9. long process of incus
10. stapes
11. posterior malleolar fold
12. short process of incus
13. body of incus
14. malleoincudal joint

No. 6-18 THE MIDDLE EAR: THE TYMPANIC CAVITY
Text pages 442–445

1. The tympanic cavity is within the _____ portion of the _____ bone.

    1. _____
    _____

2. To visualize the comparative levels of the ear canal and the tympanic cavity, note that their floors are on somewhat the same level. While the tympanic cavity does not have a basement, it does have a/an _____ or _____ recess.

    2. _____
    _____

3. The portion of the tympanic cavity whose lateral border is the drum membrane is called the tympanic cavity _____.

    3. _____

4. The tympanic aditus, an orifice on the *anterior / posterior* wall of the attic, provides an opening into the tympanic antrum which communicates with the _____ cells.

    4. _____

    Why can an infection in the middle ear cavity readily spread to these cells?

    _____

5. The tegmen tympanum, a paper-thin plate of bone forming the roof of the tympanic cavity and antrum, separates the tympanum from the cranium and the _____.

    5. _____

6. The tympanic plate of the temporal bone forms the floor of the tympanic cavity, separating it from the groove housing the _____ vein.

    6. _____

7. The lateral wall of the epitympanic recess is *membranous / cartilaginous / bony*.

    7. _____

8. A small branch of the facial nerve which courses in and out of the lateral portion of the tympanic cavity is known as the _____ tympani.

    8. _____

9. The septum canalis musculotubarii separates the somewhat parallel canals for the tensor tympani and the _____ tube.

    9. _____

Description/Landmarks Walls of Tymp. Cavity:	Lateral	Anterior	Medial	Posterior
membranous wall	✔			
mastoid wall				✔
labyrinthian wall			✔	
carotid wall		✔		
formed primarily by petrous portion of temporal bone		✔	✔	✔
formed by drum membrane and in part by squamous portion of temporal bone	✔			
oval window			✔	
perforated by tendon for tensor tympani m.		✔		
round window			✔	
entrance to tympanic antrum				✔
aperture of auditory (Eustachian) tube		✔		
pyramidal eminence containing stapedius m.				✔
fossa incudis housing short process of incus				✔

Description	Oval Window	Round Window
opening into the vestibule of the inner ear	✔	
opening into the scala tympani of the cochlea		✔
located superiorly	✔	
occupied by the footplate of the stapes	✔	
closed by the secondary tympanic membrane		✔

FIGURE 6.5 SCHEMATIC MIDDLE EAR AS SEEN FROM THE FRONT.

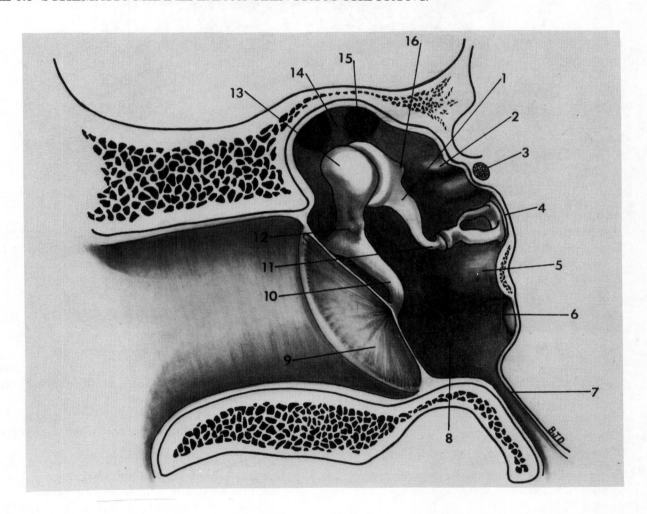

Identify:

1. long process of incus
2. prominence of lateral semicircular canal
3. facial nerve
4. oval window
5. promontory

6. round window
7. auditory (Eustachian) tube
8. middle ear cavity
9. tympanic membrane
10. manubrium of malleus

11. lenticular process of incus
12. anterior process of malleus
13. head of malleus
14. epitympanic recess
15. aditus to tympanic antrum
16. short process of incus

FIGURE 6.6 SCHEMATIC CORONAL SECTION OF THE HUMAN EAR.

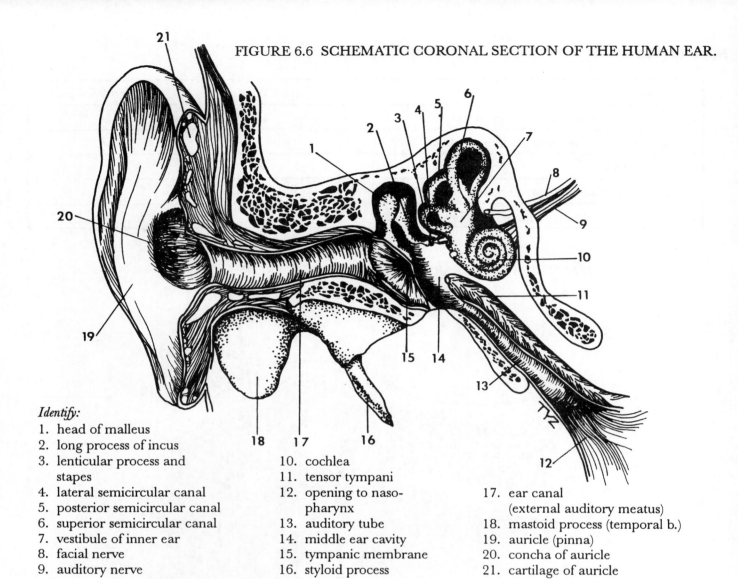

Identify:

1. head of malleus
2. long process of incus
3. lenticular process and stapes
4. lateral semicircular canal
5. posterior semicircular canal
6. superior semicircular canal
7. vestibule of inner ear
8. facial nerve
9. auditory nerve
10. cochlea
11. tensor tympani
12. opening to naso-pharynx
13. auditory tube
14. middle ear cavity
15. tympanic membrane
16. styloid process
17. ear canal (external auditory meatus)
18. mastoid process (temporal b.)
19. auricle (pinna)
20. concha of auricle
21. cartilage of auricle

Note: Disorders of the Auditory Tube:

<u>patent (open, expanded)</u> may be the result of extreme weight loss

Symptoms:

1. sensation of fullness in the ear
2. bothered by hearing one's own breathing
3. tympanic membrane moves in and out as person breathes
4. one's own voice seems very loud
5. symptoms occur only when person is upright

<u>blocked or closed</u> may be the result of enlarged adenoids

1. occasional mild earache
2. mild fluctuating hearing loss
3. sensation of fullness in the ears
4. drum membrane may be retracted
5. may have fluid in tympanic cavity

Note: Middle ear infections seem to be most common in children from six months to two years old. In about 10% the tympanic membrane will rupture, but scarring does not seem to affect function. About 80% of the infections will clear up without medication, although effusion (collection of fluid in the tympanic cavity) may last up to twelve weeks. Some studies indicate that hypofunction of the Eustachian tube may predispose some children to the development of chronic otitis media with effusion. Myringotomy (incision of the tympanic membrane) and insertion of tympanostomy (ventilating) tubes has been a relatively common treatment for persistent middle ear problems, but its usage has declined.

Description Divisions of Tube:	Osseous	Cartilaginous	Membranous	Isthmus
forms tympanic opening	✔			
junctions of osseous and cartilaginous portions				✔
forms pharyngeal opening		✔		
contributes to torus tubarius		✔		
lining of the tube			✔	
normally collapsed, but frequently dilated during swallowing		✔		

1. The auditory tube establishes communication between the _____ and the _____.

   1. _____

2. Its length is approximately 35-38 mm or *1.3 / 2.3 / 3.3* inches.

   2. _____

3. As the auditory tube leaves the tympanic cavity, in what directions does it course? In children the course is more _____.

   3. _____
   _____

   Why is this significant? _____

4. What are the functions of the auditory tube?

   4. _____
   _____

5. A discharge of a fluid tissue is called _____.

   5. _____

   Explain why this fluid may partially fill the tympanic cavity even when the auditory tube is functioning properly.

   _____
   _____

6. In children, the pharyngeal opening of the auditory tube is sometimes blocked by the _____.

   6. _____

7. If the auditory tube is closed (chronically) the air in the middle ear cavity will be absorbed by the _____, and the pressure within the cavity will become *positive / negative*.

   7. _____
   _____

8. The middle ear functions best when its pressure differential with respect to ambient air pressure is *positive / negative / zero*.

   8. _____

   Explain why. _____

Quote: ". . . the tube that connects the ear space to the mouth (the Eustachian tube) is quite floppy in the young child. The best analogy are the old paper straws we use to drink milk with when we were in kindergarten; remember how floppy and collapsed the paper straw would get when it became wet? A child's Eustachian tube is much like that (the technical term is patulence) and this makes the drainage of fluid from ear space to mouth difficult, leading to a build up of fluid behind the ear drum and a risk of ear infection."

H. Markel and F. Oski, 1996

No. 6-20 THE MIDDLE EAR: THE AUDITORY OSSICLES
Text pages 448–452

Description	Malleus	Incus	Stapes
resembles sculptor's mallet	✔		
resembles anvil		✔	
resembles stirrup			✔
head occupies one-half of attic (epitympanic recess)	✔		
first ossicle (most lateral)	✔		
second ossicle		✔	
third ossicle (most medial)			✔
smallest			✔
attached to tympanic membrane	✔		
consists of a body and two crura (arms)		✔	
consists of a head, neck, two crura, and a footplate			✔
consists of a head, neck, and three processes	✔		
portion occupies oval window			✔
encloses space called obturator foramen			✔

1. When do the ossicles achieve their full growth? 1. _____

2. What are the two primary 2. _____
   functions of the ossicles? _____

3. The footplate of the stapes is fastened to the bony walls of   3. _____
   the oval window by the _____ ligament.

4. Articulated by the malleoincudal joint, the _____ and   4. _____
   _____ move as a single mass *even / except* when driven   _____
   by powerful stimuli.   _____

5. The incus and stapes are articulated by a flexible ball-and-   5. _____
   socket joint known as the _____ joint.

6. If the ossicular chain were to continue vibrating after the   6. _____
   sound vibrations had ceased, this would be a potential
   source of _____ in the middle ear.

7. The ligamentous system suspending the ossicular chain is   7.   T   F
   primarily responsible for this cessation. Explain your answer.

   _____

   _____

   _____

8. If you were to see the inside of an actual tympanic   8. _____
   cavity with all its structures in place, you would
   probably have difficulty identifying the ossicles. Why?   _____

FIGURE 6.7 SCHEMATIC ILLUSTRATION OF MIDDLE EAR LIGAMENTS AND THE STAPEDIUS MUSCLE.

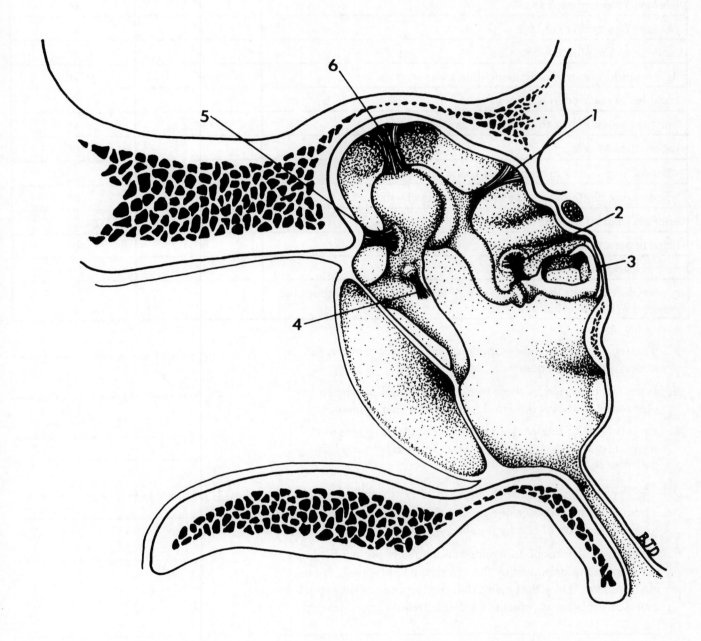

Identify:

1. posterior ligament of incus
2. stapedius muscle
3. annular ligament of stapes
4. stump of anterior malleolar ligament
5. lateral malleolar ligament
6. superior malleolar ligament

Description	Tensor Tympani	Stapedius
primarily striated muscle	✔	✔
pennate (featherlike) muscle	✔	✔
can exert much force for its size	✔	✔
its tendon has more elastic tissue than most tendons	✔	✔
completely encased in a bony canal	✔	✔
only its tendon enters the tympanic cavity	✔	✔
the smaller muscle		✔
when acting alone functions as its name implies	✔	
when acting alone pulls stapes away from oval window		✔
appears to exert greater force	✔	
helps fixate the ossicular chain	✔	✔
more easily fatigued	✔	
contraction permits the transmission of vibratory energy without the generation of distortion	✔	✔

1. The tensor tympani and stapedius muscles exert force in the *same / opposite* direction.

   1. _____

2. Is the force exerted by these muscles horizontal or perpendicular to the primary rotational axis of the ossicular chain?

   2. _____

3. An involuntary, relatively invariable adaptive response to a stimulus is a/an _____. If the stimulus is sound it is called a/an _____.

   3. _____
   _____

4. The time lag between stimulus and response is called _____.

   4. _____

5. The apparent resistance of a mechanical or electrical system to the absorption of energy is called _____.

   5. _____

6. The tympanic musculature cannot adequately protect the inner ear from being damaged by the extremely loud sounds in our environment. Does that mean that the response of the tympanic musculature is not a protective reflex? Defend your answer.

   6. _____

   _____

   _____

7. Morphologically the tensor tympani and stapedius muscles are *synergists / antagonists*, but physiologically they are *synergists / antagonists*.

   7. _____
   _____

8. Another name for the protective function of tympanic musculature is _____ control.

   8. _____

9. The protective function is limited by the latency of _____.

   9. _____

10. The Acoustic Reflex:

    a. may be elicited by any sound

    b. elicited by sounds approximately 80 dB above threshold for hearing

    c. reduction in sound transmission greater for high frequency sounds

    d. muscle contraction begins at the same time the sound begins

    e. muscle contraction continues until the sound ends

    f. decreases acoustical impedance of ossicular chain

10.

    a. T F

    b. T F

    c. T F

    d. T F

    e. T F

    f. T F

11. If the tympanic muscles are fatigued by prolonged exposure to intense sound they will respond to

    a. increased intensity at the same frequency.

    b. the same intensity at a different frequency.

    c. both

11. _____ a _____ b _____ c _____

12. As intensity increases, a tilting action is added to the usual pistonlike movement of the _____.

12. _____

No. 6-22 TRANSFORMER ACTION OF THE MIDDLE EAR
Text pages 458–464

1. What happens when sound waves traveling in one medium encounter a different medium?

1. _____
_____

2. The acoustic resistance of a medium is determined by its _____ and _____.

2. _____
_____

3. The structures of the middle ear act as a/an _____ matching device, equating the resistance of the load to that of the source.

3. _____

4.

Description	Tympanic Membrane	Oval Window
a. pressure at this membrane is the source		
b. pressure at this membrane is the load		
c. moves through little distance with great force		
d. moves through great distance with comparatively little force		

5. What are three ways in which airborne sounds can be transmitted to the inner ear?

5. _____
_____
_____

6. If the ossicles are removed the airborne sounds exert pressure on both the oval window and round window.

   6.

  a. The airborne hearing loss will be about *10 / 30 / 60* dB.

   a. _____

  b. How is the basilar membrane affected by this pressure on both windows?

   b. _____

  c. if the round window is occluded or shielded the loss will be about _____ dB.

   c. _____

7. The tympanic membrane vibrates like a solid disc pivoted on an axis when it is stimulated by *low / high* frequency sounds, with the greater amplitude of vibration occurring at the *upper / lower* edge of the membrane.

   7. _____

8. The drum vibrates segmentally at *low / high* frequencies.

   8. _____

9. Which of the following appears to make the greatest contribution to the transformer action of the middle ear?

   9. _____ a   b   c

  a. the mode of vibration of the drum membrane

  b. the mechanical advantage provided by the ossicular chain

  c. the mechanical advantage provided by the differences in effective areas of the drum membrane and the stapes footplate

10. Why is the effective area of the drum membrane smaller than its actual area?

   10. _____

11. Force per unit area = _____.

   11. _____

12. Pressure × area = _____.

   12. _____

13. The forces exerted on the drum membrane and the stapes footplate are *the same / different*.

   13. _____

14. The pressures exerted on the drum membrane and the stapes footplate are *the same / different*.

   14. _____

15. Pressure transformation decreases above *500 / 2500 / 5000* Hz.

   15. _____

Note: Pressure at stapes footplate $= \dfrac{\text{force at ear drum}}{\text{stapedial footplate area}} = \dfrac{\text{pressure} \times \text{area of ear drum}}{\text{stapedial footplate area}}$

Pressure at the ear drum $= \dfrac{\text{force acting on drum membrane}}{\text{effective area of drum membrane}}$

No. 6-23 THE INNER EAR: THE BONY LABYRINTH
Text pages 464–467

Description Bony Labyrinth:	Vestibule	Semicircular Canals	Cochlea
part of the otic capsule	✔	✔	✔
canals or cavities within the petrous portion of temporal bone	✔	✔	✔
attains full size before birth	✔	✔	✔
coiled around modiolus (central pillar of bone)			✔
three in number		✔	
perforated by round window			✔
perforated by oval window	✔		
perforated by cochlear aqueduct			✔
five orifices opening into vestibule		✔	
scala vestibuli is upper duct			✔
scale tympani is lower duct			✔
divided by osseous spiral lamina (bony shelf)			✔
helicotrema establishes communication between its two ducts			✔
lateral wall forms vestibular wall of middle ear cavity	✔		
central portion of labyrinth	✔		
lined with a thin, fibroserous membrane	✔	✔	✔
free surface of lining covered with perilymph-secreting epithelium	✔	✔	✔

1. Functionally the inner ear is divided into two cavity
    systems, one housing the organs of _____ and the
    other the essential organ of _____.

1. _____
   _____

2. Structurally the inner ear is divided into two labyrinthine
    systems, the _____ labyrinth contained within the
    _____ labyrinth.

2. _____
   _____

3. Like cerebrospinal fluid, perilymph is an ultrafiltrate of
    _____, and its ionic composition is similar to that of
    intracellular / extracellular fluid. Endolymph, which fills
    the membranous labyrinth, has an ionic composition similar to
    intracellular / extracellular fluid.

3. _____
   _____
   _____

4. Perilymph fills the *scala vestibuli / scala tympani / both*.

4. _____

5. Perilymph also fills the perilymphatic spaces around the
    semicircular canals and within the _____.

5. _____

FIGURE 6.8 THE BONY LABYRINTH.

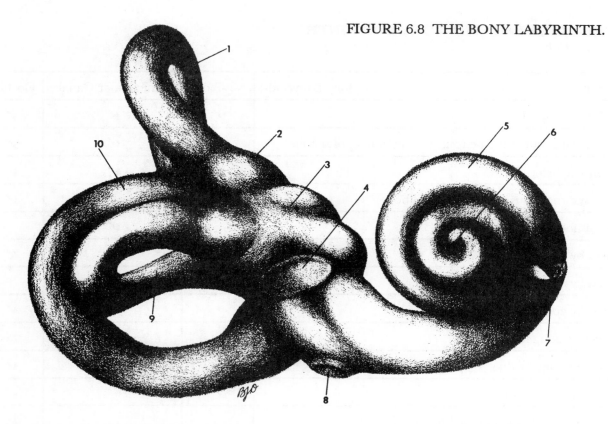

Identify: 1. superior semicircular canal

2. ampulla

3. vestibule

4. oval window

5. cochlea

6. cupola of cochlea

7. ductus cochlearis

8. round window

9. lateral semicircular canal

10. posterior semicircular canal

No. 6-24 THE INNER EAR: THE MEMBRANOUS LABYRINTH
Text pages 467–469

Description	Semicircular Canals	Utricle and Saccule	Cochlear Duct (Scala Media)
division of the membranous labyrinth	✔	✔	✔
comprises the system of hearing			✔
part of the system for equilibrium	✔	✔	
contained within the vestibule		✔	
united by endolymphatic duct		✔	
filled with endolymph (a viscous ultrafiltrate of blood)	✔	✔	✔
within the bony labyrinth	✔	✔	✔

1. To what do sensory organs in the ampullae, utricle, and saccule respond?

1. _____

_____

No. 6-24 cont'd

2. The division of the equilibrium system which functions in perception of position in space and the vertical plane is called the *static / kinetic* system.

2. _____

3. The division of the equilibrium system which functions in perception of rotation and acceleration of the head is the *static / kinetic* system.

3. _____

No. 6-25 THE MEMBRANOUS LABYRINTH: THE COCHLEAR DUCT (SCALA MEDIA)
Text pages 469–471

Schematic of a cross-section through the cochlea (without spiral organ).

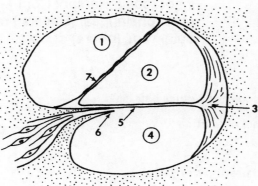

Identify:

(1) scala vestibuli

(2) cochlear duct (scala media)

(3) spiral ligament

(4) scala tympani

(5) basilar membrane

(6) spiral lamina

(7) vestibular membrane

	Description	Name or Number
1.	contains perilymph (2)	
2.	contains endolymph	
3.	directly communicates with vestibule	
4.	communicates through the helicotrema (2)	
5.	supports the essential organ of hearing	
6.	forms floor of scala media, roof of scala tympani (2)	
7.	separates scala vestibuli and scala media	
8.	narrow shelf of bony plates	
9.	thickening of periosteum projecting inward to form basilar crest	
10.	houses canals for peripheral fibers of auditory nerve	
11.	when viewed from above resembles "washboard" road	

Question: What is the only outlet of the cochlear duct?

No. 6-26 THE INNER EAR: THE SPIRAL ORGAN (OF CORTI)
Text pages 471–477 *Study figures in text.*

inner (with respect to modiolus)						outer
1st	2nd	3rd	4th	5th	6th	7th

border cells
of Held

Arrangement of cells on the spiral lamina and basilar membrane.

Write the names of the supportive cells in their proper order as demonstrated above. Underline the names of the cells that most directly support the receptive cells (the inner and outer hair cells).

border cells of Held   inner rods (pillars) of Corti
cells of Claudius   outer phalangeal (Deiters') cells
cells of Henson   outer rods (pillars) of Corti
inner phalangeal cells

1. Which cells form cups to accommodate the outer hair cells?   1. _____

2. Which cells enclose the inner tunnel of Corti?   2. _____
_____

3. What membrane is formed by the processes or plates of cells 2–5?   3. _____

4. Which secretory cells contribute to the formation of the stria vascularis?   4. _____

5. Which secretory cells partially line the internal spiral sulcus?   5. _____

6. There are three or four rows of *inner / outer* hair cells and one row of *inner / outer* hair cells.   6. _____
_____

7. The area of the bony spiral lamina pierced with openings for nerve fibers entering the organ of Corti is the habenula _____ .   7. _____

8. The base of a hair cell is supported by a _____ cell, while its apex is supported by a _____ process.   8. _____
_____

9. On the side walls of the inner tunnel of Corti there are openings between the rods (pillars) which permit circulation of *perilymph / endolymph* and allow passage of _____ fibers. *The fluid is sometimes called cortilymph.*   9. _____
_____

10. Hair cells occupy spaces in the netlike *reticular / tectorial* membrane and project toward the rooflike *reticular / tectorial* membrane.   10. _____
_____

11. The stria vascularis which lines the outermost wall of the scala _____ probably secretes *perilymph / endolymph.*   11. _____
_____

12. All of the cells resting on the basilar membrane are *supportive / receptive.* The inner and outer hair cells are the only _____ cells.   12. _____
_____

Description Hair Cells:	Inner	Outer
number 12,000 to 20,000		✔
number approximately 3,500	✔	
supported by phalangeal cells	✔	✔
innervated by nerve fibers coursing through the habenula perforata	✔	✔
entire cell including stereocilia surrounded by plasma membrane	✔	✔
outer row of stereocilia is the longest	✔	✔
coarser stereocilia	✔	
stereocilia distributed in the form of a W		✔
probably respond primarily to a lateral (radial) shearing movement	✔	✔

13. The proper term for the hairs of the hair cells is _____.

    13. _____

14. Because the tectorial membrane is completely noncellular, its functions seem to be purely *metabolic / bioelectrical / mechanical.*

    14. _____

15. The layers of the tectorial membrane:

    a. top

    b. middle

    c. bottom

    15.

    a. _____

    b. _____

    c. _____

16. The contact of the stereocilia with the tectorial membrane is probably more extensive *prenatally / postnatally.*

    Why? _____

    16. _____

17. Small strands of fiber maintain contact between the tectorial membrane and _____ cells.

    17. _____

No. 6-27 THE FUNCTION OF THE INNER EAR: THEORIES OF HEARING
Text pages 478–482 *Volley theory covered in 6-29.*

1. Theories of hearing deal with the transformation of vibratory energy into _____ impulses.

   1. _____

2. Because the membranous labyrinth is a/an *closed / open* system contained within a/an *yielding / unyielding* labyrinth, and the labyrinthine fluids *are / are not* readily compressed, it is difficult to determine exactly how vibration of the stapes footplate causes displacement of the cochlear _____.

   2. _____

      _____

      _____

      _____

3. Describe two theories concerning   3. _____
   the mechanics of fluid movements
   within the cochlea.   _____

4.

Subclasses	Theories of Hearing:	Place	Frequency
a.	telephone or nonanalytic		
b.	traveling wave or nonresonance		
c.	frequency analytic		
d.	resonance		

5. According to the resonance theory formulated by Helmholtz,   5. _____
   every fiber in the auditory nerve was associated with a
   different _____ .

6. What element of the resonance theory has   6. _____
   been refuted by research on the physical
   properties of the basilar membrane?   _____

7. The fibers in the basilar membrane which prevent wave propaga-   7. _____
   tion in the longitudinal direction are *radial / longitudinal.* Groups
   of these fibers, being characteristically independent, have   _____
   relatively *unique / common* response characteristics to oscillating pressure gradients within the cochlear fluids.

8. Although Békésy found that the basilar membrane was not under   8. _____
   appreciable tension, he noted much greater stiffness at the
   narrow *basal / apical* end.

9. In contrast to the place theories, the nonanalytic frequency   9. _____
   theory does not endow the cochlea with any _____ function.

10. A device that absorbs energy in one form and emits energy,   10. _____
    usually in a different form, is called a/an _____ .

11. The nonanalytic frequency theory regards the inner ear as   11. _____
    a/an _____ which transforms vibratory energy into
    coded patterns of _____ impulses.   _____

12. According to the nonanalytic frequency theory, analysis and   12. _____
    discrimination occur in the _____ .

13. The basis of nonanalytic frequency theory resembles that of   13. _____
    the _____ .

14. What did nonanalytic frequency theories   14. _____
    assume about the auditory nerve fibers?
                                                   _____

15. What are the actual differences between   15. _____
    auditory nerve fibers and other sensory
    fibers?

16. The study of noise-induced damage to the inner ear has   16. _____
    tended to *support / refute* the nonanalytic frequency theory.

    Explain. _____

    _____

17. The nonanalytic frequency theories would   17. _____
    require that the basilar membrane move
    up and down in its entirety. Why doesn't   _____
    this type of movement take place?

18. The standing wave theory has been refuted, in part,
    because Ewald was using incorrect values for compliance
    and resistance of the basilar membrane.

    For what other reasons has   18. _____
    it been refuted?   _____

19. The traveling wave theories assumed the spatial distribution   19. _____
    of basilar membrane displacement was related to _____.   _____
    Research has shown that high tones stimulate the *basal /*
    apical end of the cochlea, and low tones the *basal / apical* end.   _____

20. Traveling wave theories had difficulty reconciling the   20. _____
    distance and velocity of motion generated by the inward
    and outward movement of the _____.

21. Békésy's construct of cochlear hydrodynamics is usually   21. _____
    classified as a _____ theory.

    Why may the name of the theory   _____
    be somewhat misleading?   _____

22. According to Pascal's principle, pressure at any point in a   22. _____
    closed-fluid system is transmitted to all other points in
    the system. Because cochlear fluids are virtually imcom-
    pressible, pressure changes will be *gradually / instantaneously*
    transmitted throughout the cochlea.

23. While perilymph is a watery liquid, endolymph has the vis-   23. _____
    cosity of _____. This difference in physical proper-
    ties constitutes a/an _____ (a surface which is the   _____
    common boundary between two parts or spaces).

24. As longitudinal surface waves generated at the _____   24. _____
    travel through the perilymph, they create periodic pressure
    patterns which change in _____ and _____ across the   _____
    cochlear partition.

    These pressures are transmitted across the cochlear parti-   _____
    tion to the _____ media, and ultimately reach the round
    window which then _____.   _____

25. Why is this release of pressure   25. _____
    at the round window so important?   _____

Note: The cochlear partition includes the basilar and tectorial membranes, supportive and receptor cells, and endolymph.

1. The displacement pattern of the basilar membrane is due to its _____ characteristics.

2. The membrane is about *0.1 / 0.5* mm wide at the base, and *0.1 / 0.5* mm wide at the apex.

3. Stiffness increases approximately 100-fold from the _____ to the _____.

4. Does the coupling of the segments along the membrane contribute to the response patterns of the membrane?

5. Displacement patterns on the basilar membrane.

1. _____

2. _____

   _____

3. _____

   _____

4. _____

Identify:

High frequency pattern	A	B	C
Mid-frequency pattern	A	B	C
Low frequency pattern	A	B	C

Label (just once):

malleus

stapes

round window

oval window

helicotrema

scala media

scala tympani

tympanic membrane

auditory (Eustachian) tube

semicircular canal

basilar membrane extremities:

    low frequency

    high frequency

    wide

    narrow

    stiff

    flaccid

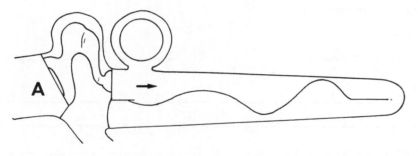

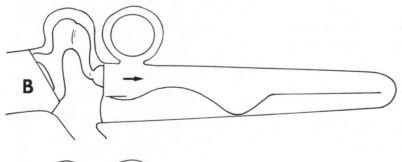

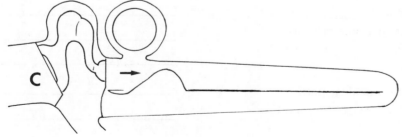

6. Features of the Traveling Wave on the Basilar Membrane.   6.

   a. Locus of maximum disturbance on the membrane is   a. _____
      _____ dependent.

   b. Just beyond the point of maximum displacement, ampli-   b. _____
      tude quickly reaches _____. Because of the inter-   _____
      action and energy exchange between the cochlear parti-
      tion and the perilymph at the interface, the amplitude
      of the waves in the perilymphatic scalae replicate those
      of the _____.

   c. Why do low frequencies tend   c. _____
      to mask high frequencies?   _____

   d. The velocity and therefore the wavelength decrease as   d. _____
      the distance from the stapes *increases / decreases.*

   e. Theoretically, the point of excitation *will / will not /*   e. _____
      will in part determine the location or extent of the
      disturbance on the membrane.

   f. Multiple regions of maximum disturbance will be produced   f. _____
      by _____ stimuli.

   g. Intensity influences the *location / amplitude / location and*   g. _____
      amplitude of the maximum displacement.

   h. The mechanical frequency analysis performed by the cochlea   h. _____
      supports a *place / frequency* theory of hearing.

   i. The most influential physical property of the basilar   i. _____
      membrane appears to be its changes in *width / stiffness.*

   j. The pronounced peak of a traveling   j. _____
      wave may be nonlinear. Why?   _____

   k. What was proven by Békésy's many attempts to modify the   k. _____
      vibratory pattern without modifying the characteristics
      of the membrane itself?

No. 6-29 ANALYTIC THEORY; THE VOLLEY PRINCIPLE
Text pages 486–487

1. Auditory nerve discharges are synchronous with auditory stimulation up to approximately *1,000 / 4,000 / 8,000* per second, and then gradually give way to asynchronous nerve discharges to as high as *8,000 / 12,000 / 15,000* per second.

   1. _____

      _____

2. Can the frequency of nerve discharges represent pitch throughout the entire hearing range?

   2. _____

3. The alternate firing of individual nerves to increase the response rate of the auditory system is known as the _____ principle.

   3. _____

4. The volley theory combines _____ and _____ theories.

   4. _____

      _____

5. In the volley theory, stimulus intensity is represented by what two factors?

   5. _____

      _____

6. Do the neurons leading from the basilar membrane retain their frequency identity to the level of the cerebral cortex?

   6. _____

Volley Theory	Contributing Theories:	Place	Frequency
pitch analysis		✔	✔
15–400 Hz			✔
400–5,000 Hz		✔	✔
5,000–20,000 Hz		✔	

No. 6-30 EXCITATION OF THE HAIR CELLS
Text pages 487–489

Quotes: "Viscosity—the physical property of a fluid or semifluid that enables it to develop and maintain a certain amount of shearing stress dependent upon the velocity of the flow and then to offer continued resistance to flow . . ."

Webster's Third New International Dictionary, 1986

"Shear. A force which lies in the plane of an area or a parallel plane is called a shearing force. It is the force which tends to cause the plane of the area to slide on the adjacent planes."

Van Nostrand's Scientific Encyclopedia, 1989

1. When a membrane or plate is bent into a curve, its inner edge will be *compressed / stretched*, and its outer edge _____.

   What kind of force will be generated between the inside and outside radius?

2. Equal vertical displacement of the tectorial and basilar membranes will produce shearing forces because the membranes are hinged at *different / the same* points. Would the force of the vertical displacement or the force of the shear be of greater magnitude?

3. Only the outermost row of the stereocilia of the outer hair cells seem to make contact with the _____ membrane.

4. During movement of the cochlear partition, the free-standing stereocilia are probably stimulated by the viscous drag of the streaming *perilymph / endolymph*.

5. The bending of the outer row of stereocilia of the outer hair cells is proportional to _____ of the cochlear partition, while the bending of the free-standing outer and inner hair cell stereocilia is proportional to the *velocity / amplitude* of basilar membrane movement.

6. It has been theorized that deformation of the stereocilia may result in a change in the _____ resistance of the hair cell thus generating a receptor current which flows through the _____.

   The presence of differing concentrations of potassium and sodium in the endolymph and perilymph would tend to *refute / support* this theory.

1. _____
   _____
   _____

2. _____
   _____

3. _____

4. _____

5. _____
   _____

6. _____
   _____
   _____

No. 6-31 THE NEUROPHYSIOLOGY OF THE COCHLEA
Text pages 489–494 *Review No. 5-29, page 217.*

1. Neurophysiologically the cochlea, as it converts acoustic (mechanical) energy into nerve impulses and other bio-electric potentials, behaves as a/an *amplifier / tranducer*.

2. The initial disturbance of the hair cell membrane and the resulting polarization are the result of shearing forces acting on the _____ and of viscous streaming of the *endolymph / perilymph*.

3. Resting Potentials:

   a. Acoustic stimuli only trigger the process of local energy conversion to neural impulses.

   b. The endocochlear potential is not affected by destruction of the hair cells.

   c. The endocochlear potential is not oxygen dependent.

   d. Intracellular negative membrane potential of the hair cells probably originates from the cortilymph.

1. _____

2. _____
   _____

3.

   a. T F

   b. T F

   c. T F

   d. T F

4. Summating Potentials:

    a. Summating potentials, produced by placing electrodes in the scala _____ and the scala _____, are evoked by a/an _____ stimulus.

    b. Although summating potentials were identified years ago, they are not clearly understood because they are composites of a number of bioelectric components which are not easily *stimulated / separated.*

    c. Summating potentials, thought to be primarily a product of the *outer / inner* hair cells, are closely related to cochlear _____.

5. Cochlear Microphonics:

    a. As the term cochlear microphonics suggests, when an electrode is placed in the vicinity of the cochlea an electric _____ will accurately reproduce the frequency and wave form of the sound stimulus. The cochlea seems to function as a biological _____.

    b. Cochlear microphonics seem to be generated by the *hair cells / auditory nerve / both.*

    c. The amplitude of cochlear microphonics is proportional to that of the stimulus except at very *high / low* sound pressure levels.

    d. Comment on the following aspects of cochlear microphonics:

        distortion

        threshold

        adaptation to stimulus

        fatigue

        frequency limits

    e. Although cochlear microphonics may be only an "epiphenomenon," what information may be gained by their use?

    f. The technique of establishing time-space patterns of microphonics from different points on the cochlea is sometimes called _____ mapping.

    g. What is another method of generating cochlear maps?

    h. Information derived from cochlear frequency mapping:

        (1) Low frequencies are crowded into _____ region.

        (2) High frequencies are spread out over _____ region.

        (3) More diffuse damage results from exposure to intense *high / low* frequency sounds.

        (4) Results support _____ principle.

        (5) Results support *place / frequency* theories of hearing.

4.

    a. _____

       _____

       _____

    b. _____

    c. _____

       _____

5.

    a. _____

       _____

    b. _____

    c. _____

    e. _____

       _____

    f. _____

    g. _____

       _____

    h.

       (1) _____

       (2) _____

       (3) _____

       (4) _____

       (5) _____

6. Action Potentials (Whole Nerve Potentials):

   a. Action potentials can be recorded from the _____ or the trunk of the _____ nerve.

   b. Are action potentials true cochlear potentials?

   c. Action potentials generated by complex stimuli are asynchronous conglomerate _____.

   d. What stimuli are usually used to elicit whole nerve action potentials?

   e. Discharges become more asynchronous as the frequency level of the stimulus is *raised / lowered.*

6.

   a. _____

      _____

   b. _____

   c. _____

   d. _____

      _____

   e. _____

Description Potentials:	Resting	Cochlear Microphonics	Summating	Action
endocochlear / intracellular potentials	✔			
stimulus-related potential		✔	✔	✔
exists without acoustic stimulus	✔			
composite of bioelectric components			✔	
mimic sound stimulus		✔		
closely reflect mechanical events occurring on basilar membrane		✔		
help determine functional integrity of inner ear		✔		
membrane potential	✔			
whole nerve potential				✔
magnitude of potential does not seem to reach saturation level		✔	✔	
highly dependent on blood supply to inner ear		✔		
useful for tonotopographical mapping		✔		
generated by cochlear structures		✔	✔	
generated by auditory neurons				✔

No. 6-32 THE EVOKED COCHLEAR MECHANICAL RESPONSE
Text page 494

The evoked cochlear mechanical response is

1. a resting neural potential.

2. an active neural potential.

3. apparently the result of reflection.

4. similar to an echo, but is far below the threshold of hearing.

5. found in persons with sensorineural deafness.

1. T F

2. T F

3. T F

4. T F

5. T F

No. 6-33 TRANSDUCTION IN THE COCHLEA
Text pages 494–499

1. Displacement patterns on the basilar membrane are _____ specific.

2. Bending of the stereocilia, by shearing forces or by viscous streaming, results in what kind of changes in the hair cell?

   a. These potentials are conducted to the region of the _____ perforata.

   b. Beginning in this region, the nerve fiber is *myelinated / nonmyelinated.*

   c. The spike potential is generated at the first _____ of Ranvier.

   d. Neural impulses eventually reach the _____.

3. The energy contributed by the spiral organ assists in transforming acoustic (fluid pressure gradients) energy into _____ impulses.

4. If polarization of the scala media is increased

   a. the cochlear microphonic is *increased / decreased.*

   b. the summating potential is *increased / decreased.*

5. Bending (shearing) of the hair cells:

   a. The amount of current flow is dependent on the *degree / direction / degree and direction* of shear.

   b. Alternating to and fro movement of the cilia yields *cochlear microphonics / summating potentials.*

   c. A steady one-directional shear yields _____.

6. The graded potential at the hair cell is called the local or *receptor / resting* potential.

7. The graded neural response of the peripheral unmyelinated part of the nerve fiber is called the _____ potential because it initiates the _____ action potential.

1. _____

2. _____

a. _____

b. _____

c. _____

d. _____

3. _____

4.

a. _____

b. _____

5.

a. _____

b. _____

c. _____

6. _____

7. _____

_____

8. Responses from Inner Hair Cells:

   a. The graph of the threshold of response of a single neural unit over a range of frequencies is called a/an _____ curve.

   b. While a neuron will respond to a *wide / narrow* range of frequencies, each has its own characteristic or _____ frequency.

   c. Neural events appear to be initiated by the inner hair cells.

9. Vibration patterns on the basilar membrane

   a. If membrane displacement grows in proportion to increases in intensity the vibration pattern is *linear / nonlinear*.

   b. Current research demonstrates that the vibration pattern on the basilar membrane is very *linear / nonlinear*.

   c. In the vicinity of the frequency producing the optimum response, the displacement of the membrane is particularly *linear / nonlinear*.

   d. At frequencies above and below the optimum level the extent of the displacement is *linear / nonlinear*.

   e. With an increase in the level of sound intensity the sharpness of the tuning *increases / decreases*.

10. The outer hair cells

   a. outnumber the inner hair cells.

   b. seem to be responsible for cochlear microphonics.

   c. are candidates for a motor element on cochlear partition.

   d. are capable for storing calcium necessary for secretory/contractile activity.

   e. contain proteins associated with active contractile cells.

   f. are mechanically active; can shorten and lengthen.

   g. have rapid electromotility occurring at very high frequencies.

   h. seem to increase sensitivity of auditory nerve fibers.

   i. seem to be involved in fine-tuning of the cochlea.

11. Responses from the Auditory Nerve:

   a. The discharge rate of an unstimulated neuron is called its *normal / spontaneous* discharge rate.

      The minimal stimulus level that produces an increase of this discharge rate is its _____.

   b. As intensity increases the discharge rate _____ until it eventually levels off.

   c. Like the inner hair cells, each neuron has its own _____ frequency.

      A neuron responds better to frequencies in the *low / high* end of its range.

Answer column:

8.
  a. _____
  b. _____
  c. T F

9.
  a. _____
  b. _____
  c. _____
  d. _____
  e. _____

10.
  a. T F
  b. T F
  c. T F
  d. T F
  e. T F
  f. T F
  g. T F
  h. T F
  i. T F

11.
  a. _____
    _____
  b. _____
  c. _____
    _____

    d. When the response of a single neuron appears to have   d. _____
       a constant phase relationship to a low frequency stimulus,
       it is said to be _____.

    e. A poststimulus time histogram is a graph of _____   e. _____
       response to a *prolonged / repeated* stimulus.   _____

    f. The time between successive spike potentials is shown on   f. _____
       a/an _____ histogram.

    g. The relationship of the arrival of the neural discharge to   g. _____
       the waveform of the stimulus is shown on a/an _____
       poststimulus time histogram.

No. 6-34 THE NERVE SUPPLY TO THE COCHLEA
Text pages 499–503

  Note: Cranial Nerve VIII may be referred to as the *acoustic*, the *auditory*, or the *vestibulocochlear* nerve.

  1. Communication between the inner ear and the brain is   1. _____
     established by the _____ nerve, which divides into
     a/an _____ branch and a/an _____ branch.   _____
                                                                 _____

  2. The auditory nerve is quite *thin / thick*, and contains a   2. _____
     relatively *small / large* number of fibers.   _____

  3. It has been shown that fibers from the sympathetic branch of   3. _____
     the autonomic nervous system reach the cochlea, but do not
     enter the spiral organ. Their function in the hearing
     mechanism is probably the maintenance of _____.

Description Efferent Nerve Fibers:	Crossed	Uncrossed
constitute olivocochlear bundle	✔	✔
80 percent	✔	
20 percent		✔
arise from ipsilateral superior olivary complex		✔
arise from contralateral superior olivary complex	✔	
form inner spiral bundle		✔
tunnel radial fibers	✔	
supply outer hair cells, especially at basal turn	✔	
supply inner hair cells, more extensive apically		✔
constitute centrifugal pathway	✔	✔
stimulation generally produces inhibitory effects	✔	✔

Description Afferent Nerve Fibers:	Type I	Type II
approximately 90–95 percent	✔	
approximately 5–10 percent		✔
unmyelinated		✔
completely myelinated	✔	
peripheral (dendritic) fibers shed myelin upon entering spiral organ	✔	
supply outer hair cells		✔
radial bundles supply inner hair cells	✔	
bipolar	✔	
monopolar		✔
diffuse innervation pattern		✔
one-to-one innervation pattern	✔	
more susceptible to oxygen deficit	✔	
tonotopographical arrangement; frequency dependent sensory system	✔	
glutamate may function as hair cell neurotransmitter	✔	

FIGURE 6.9 PRINCIPAL STRUCTURES OF THE ASCENDING AUDITORY PATHWAY.

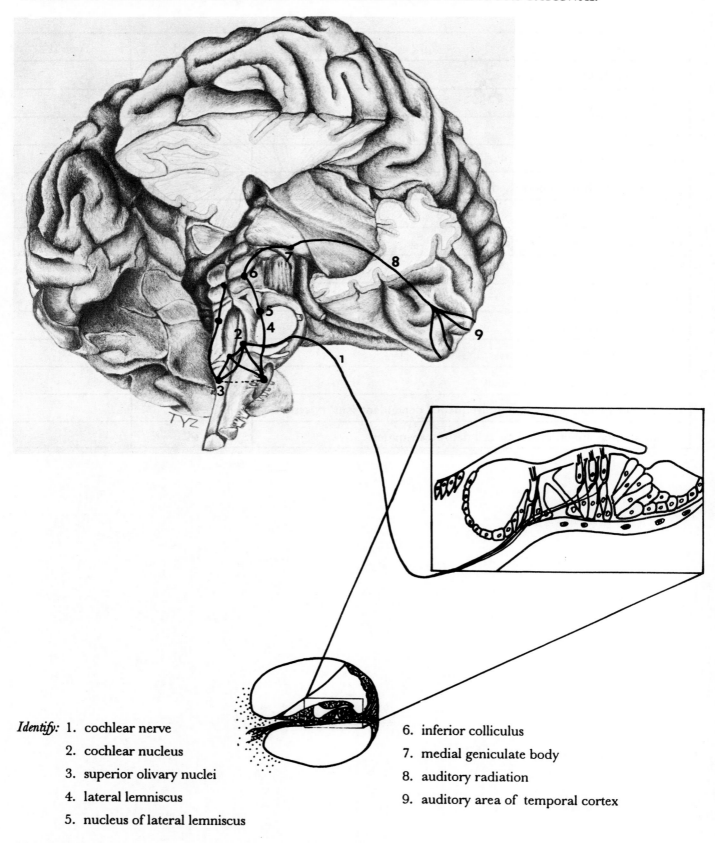

Identify:
1. cochlear nerve
2. cochlear nucleus
3. superior olivary nuclei
4. lateral lemniscus
5. nucleus of lateral lemniscus
6. inferior colliculus
7. medial geniculate body
8. auditory radiation
9. auditory area of temporal cortex

Question: At what levels do auditory nerve fibers decussate?

No. 6-35 THE ASCENDING AUDITORY PATHWAY
Text pages 503–504

1. As the auditory pathway ascends, there is a succession of at least four neurons between the cochlea and the auditory cortex of the *cerebellum / cerebrum*.

1. _____

2. The central processes of the spiral ganglion cells pass to the core of the modiolus where they form the _____ branch of the acoustic nerve.

2. _____

3. While the straight core of the nerve is formed by the most *apical / basal* fibers, the twisted, ropelike outer portion is formed by the _____ fibers.

    Why are these fibers twisted? _____

3. _____
_____

4. Because of the anatomical architecture of the cochlear nerve, the fibers most exposed and subjected to trauma are those for the *higher / lower* frequencies.

4. _____

5. In the internal auditory meatus the cochlear nerve is joined by the two divisions of the _____ nerve to form the auditory nerve.

5. _____

6. The first-order afferent neurons from the cochlea terminate in synaptic connections with the second order neurons of the dorsal and ventral _____ nuclei. Where are these nuclei located?

6. _____
_____

7. Because about half the axonal fibers of the second-order neurons decussate at the level of the cochlear nuclei, impulses from one ear reach the auditory cortex on *the same / the opposite / both* sides of the brain.

    Why is this important? _____

7. _____

8. The third-order neurons from the superior olivary complex form a tract known as the lateral _____.

8. _____

9. The inferior colliculus appears to contain centers for _____ responses to sound.

9. _____

10. From the medial geniculate body, the thalamic nucleus of the auditory pathway, fibers radiate to the _____.

10. _____

11. Since tonotopography is more diffuse at the level of the *cerebral cortex / nuclei*, frequency analysis is probably completed at the level of the _____.

11. _____
_____

12. The function of the auditory cortex may be largely *analytical / integrative*.

12. _____

1. When does vibratory energy produce   1. _____
   compressions of the skull bones?
   _____

2. What is the primary reason your   2. _____
   own voice sounds "fuller" to you
   than it does to others?   _____

3. When two tones of the same frequency are presented, one via   3. _____
   bone conduction and the other via air conduction but 180° out
   of phase, no sound will be heard. This is known as _____.

   This indicates that the vibratory patterns of the basilar   _____
   membrane produced by air- and bone-conducted sounds are
   the same / different.

4. What are the three possible avenues by which   4. _____
   bone-conducted sound can effect displacement
   of the basilar membrane?   _____

5. The basilar membrane will not be displaced if the elastic   5. _____
   characteristics of the scala vestibuli and scala tympani
   are *the same / different. See Figure 6-134 in text.*

6. Because of the greater compliance of the scala tympani,   6. _____
   generation of positive pressure displaces the basilar
   membrane into the scala _____. Generation of negative   _____
   pressure displaces the membrane into the scala _____.

7. Displacement of the skull bones produces compression of the   7. _____
   membranous _____ and its contained _____. As a
   result, additional fluid is forced into the scala _____.   _____

   _____

   Activity of the basilar membrane is thereby *diminished /*   _____
   heightened.

8. When the temporal bone is vibrating, the ossicles,   8.   T   F
   because of their inertia, remain relatively motionless.

9. Does bone conduction produced by inertial lag of the ossicu-   9. _____
   lar chain augment or diminish compression within the cochlear
   scalae?

10. When the bones of the skull are driven by a vibrator, the   10. _____
    mandible vibrates at the same _____, but is out of
    _____ with the rest of the skull.   _____

    Explain how this generates airborne   _____
    sounds within the ear canal.
    _____

_____

11. Demonstrate the occlusion effect at different pitch levels.

Vibration of Skull at Different Frequencies	200 Hz	800 Hz	1600 Hz
vibrates as a rigid body	✔		
vibrates segmentally			✔
resonates at this frequency		✔	
displacement of the skull causes compression of the fluid within the membranous labyrinth (labyrinthine bone conduction)	✔	✔	✔

Question: The occlusion effect is confined to frequencies below 2000 Hz. How might that be explained?

"We have two ears and one mouth that we may listen the more and talk the less."

Zeno, an early Greek philosopher

Chapter 7
Embryology of the
Speech and Hearing Mechanism

No. 7-1 EARLY EMBRYOLOGICAL DEVELOPMENT
Text pages 513–520

1. The paired blocks of mesodermal tissue, one block on either side of the midline, occupy the entire length of the trunk of a four-week embryo. These segments, called somites, undergo tissue differentiation resulting in cell groups called *sclerotomes*, *myotomes*, and *dermatomes*. With your knowledge of word roots, you can probably determine which term pertains to:

   a. muscle   b. bone   c. skin

1.

a. _____

b. _____

c. _____

2. When studying anatomy, how can you determine the segmental level at which a muscle originated?

   Explain your answer. _____

2. _____

_____

_____

3. Explain the origin of strong aponeurotic sheets in the body.

3. _____

_____

_____

Quote: "In man, the nerve segments which together form the neck and the arms are also the ones where the heart appears. The result is that the nerves bringing sensations from the heart are in the same segment as the nerves which bring sensation from the neck and arm. This relationship is preserved despite the fact that in the course of fetal development the heart migrates to a position which is quite remote from its original site. It sinks down through the neck into the thorax and comes to rest on the diaphragm, whose muscles are also derived from the neck segments. But the heart maintains its ancient Parliamentary representation despite its position in the body: the neck, arm, and upper chest continue to feel pain for it. . . . An ulcer on the back of the tongue may refer its pain to its old segmental partner in the ear. . . . Such pains are archaeological reminiscences of what we once were."

Jonathan Miller, 1978

No. 7-2 PRENATAL GROWTH OF THE FACIAL REGION
Text pages 520–524

1. When does flexion of the embryo occur?

2. A smooth bulge, the forebrain of the embryo, is called the *prosencephalon / mesencephalon*.

3. The transverse furrow called the oral groove (stomodeum, primitive mouth) is the topographical center of the embryonic _____.

4. Prior to its rupture and absorption during the fourth week, the buccopharyngeal membrane (oral plate) separates the foregut and the _____.

5. Arising from the anterolateral walls of the foregut, transverse elevations grow, meet at the midline, and thus form the _____ arches. The depressions between these arches are called _____ grooves or _____ clefts.

6. During the fourth week the frontonasal process is divided into a medial and two lateral nasal processes by the two _____ pits.

7. What are the four primordial areas of the face?
There are two processes and two arches.

8. The medial nasal and the maxillary processes are separated by the *oronasal / nasooptic* grooves.

9. The lateral nasal and the maxillary processes are separated by the _____ grooves.

10. The midline vertical depression of the upper lip, extending from the vermilion border to the nose is called the _____. It indicates the point of fusion between the maxillary and _____ processes.

11. The alae of the nose are formed by the *lateral / medial* nasal processes.
ala singular; *alae*, plural.   A wing, or a winglike part or process.

1. _____
2. _____
3. _____
4. _____
5. _____
   _____
6. _____
7. _____
   _____
   _____
   _____
8. _____
9. _____
10. _____
   _____
11. _____

FIGURE 7.1 THE DERIVATIVES OF THE BRANCHIAL ARCHES

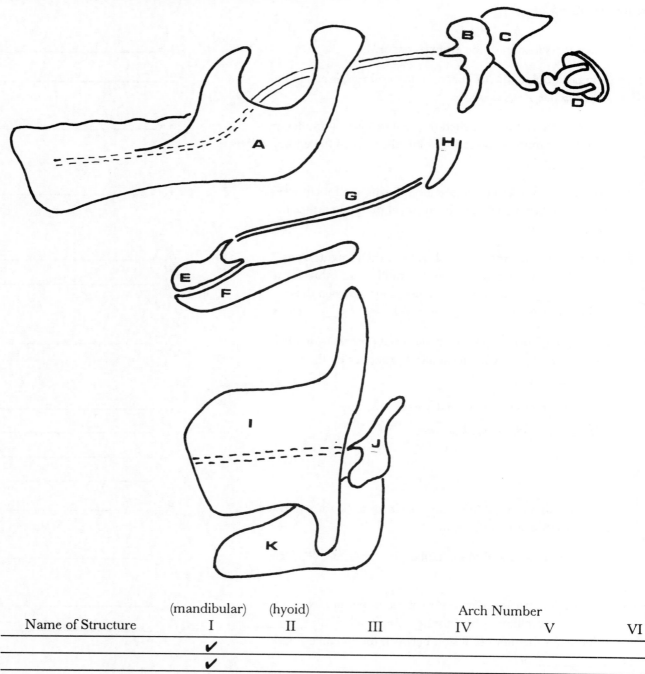

Name of Structure	(mandibular) I	(hyoid) II	III	Arch Number IV	V	VI
A.	✔					
B.	✔					
C.	✔					
D.		✔				
E.		✔				
F.			✔			
G.		✔				
H.		✔				
I.				✔		
J.					✔	
K.						∗

∗ The precise derivation of the cricoid cartilage is not known, but it probably arises from the mesenchyme of Arch VI.

1. The primary palate, a shelf of tissue separating the
    _____ and _____ cavities, begins to develop
    during the fourth week. It is formed by fusion of
    the medial nasal and _____ processes during
    the sixth week.

    1. _____

    _____

    _____

2. During the sixth week the lateral and medial nasal
    processes fuse to form the _____ (funnel-like
    openings). These openings are closed off by the
    bucconasal membrane which ruptures during seventh
    or eighth week. The openings then communicate
    directly with the _____ cavity.

    2. _____

    _____

3. During the sixth and seventh weeks the whole facial
    region grows very rapidly in a/an *anterior / posterior*
    direction.

    3. _____

4. During the eighth week the lateral areas of the mandi-
    bular arch and the maxillary processes fuse, thereby
    forming the cheeks and narrowing the _____.
    About this time the growth of the head proceeds in a
    horizontal / vertical direction.

    4. _____

    _____

5. The upper lip, the anterior portion of the alveolar
    process, and the premaxillary portion of the palate all
    develop from the primary _____.

    5. _____

6. During the eighth week the palatine processes are separ-
    ated by the _____ which at that time extends into
    the nasal cavity.

    6. _____

7. What is the result of a sudden growth
    spurt of the mandibular arch during
    the ninth week?

    7. _____

    _____

8. Later in the ninth week the palatine processes begin to
    grow in a *vertical / horizontal* direction. Fusion begins
    anteriorly / posteriorly.

    8. _____

    _____

9. The palatine processes fuse with each other and with
    the nasal _____.

    9. _____

10. The maxillary processes form the *medial / lateral* portions
    of the palate.

    10. _____

11. The palatine processes form the soft palate and *medial /
    lateral* portions of the hard palate.

    11. _____

12. The tectal ridge, a medial projection of the globular
    process, forms the _____ of the palate.

    12. _____

Quote: "Errors in facial architecture invariably take the form of clearly defined gaps or interruptions in an
otherwise normal human likeness, and since these clefts always appear in one of three or four sites and
nowhere else, one can tell that these are the natural seams."

Jonathan Miller, 1978

The primordial nervous system begins as a pair of *neural folds,* ectodermal ridges alongside the dorsal midline. Lateral to each neural fold a second ridge, the *neural crest,* is formed. Fusion of the neural folds forms the *neural tube,* soon to be covered by embryonic mesoderm. As the neural tube differentiates into the brain and spinal cord, the cells of the neural crest differentiate into ganglia of the peripheral and autonomic nervous systems.

The primitive epithelial cells comprising the early nervous system differentiate into neuroblasts (the precursors of functional neurons) and neuroglial (supportive) cells. Some neuroglial cells, the *oligodendrocytes* in the central nervous system and the *neurilemmal* cells in the peripheral nervous system, will form the myelin sheath around nerve processes.

Rapid growth of cells in the neural tube reduces the diameter of its central canal (lumen), but a longitudinal *sulcus limitans* divides its lateral wall into a *dorsal alar lamina,* which will become primarily sensory in function, and a *basal lamina* which will become motor in function. Ventral floorplates and dorsal roofplates are largely non-neural. Proliferation and differentiation of cells in the neural tube results in an internal or *ependymal zone,* a highly neural intermediate or *mantle zone,* and a relatively neuron-free *marginal zone.* Even before the rostral portion of the neural tube is completely closed three primary brain vesicles, the *prosencephalon, mesencephalon,* and *rhombencephalon,* appear. Rapid expansion of the prosencephalon results in formation of the *telencephalon,* and it encapsulates the *diencephalon.* The telencephalon and diencephalon together form the *cerebrum.* The mesencephalon remains relatively unchanged at this time, but the rhombencephalon differentiates into a rostral portion called the *metencephalon* and a caudal portion called the *myelencephalon,* which is continuous with the spinal cord. The central canal in the neural tube is also expanded in the brain region to form a system of four interconnected *ventricles,* reservoirs for cerebrospinal fluid: two large *lateral ventricles* in the telencephalon, a *third* midline *ventricle* in the region of the diencephalon, and a *fourth* in the region of the metencephalon and myelencephalon.

The myelencephalon develops into the *medulla oblongata* which contains ascending and descending nerve tracts and from which cranial nerves C V and C X will eventually emerge. The metencephalon differentiates into the *pons* (a transmission pathway between the telencephalon and cerebellum) and the *cerebellum* (an important integrating and coordinating center for body position and movement). Fusion of the two cerebellar hemispheres results in a midline structure called the *vermis.*

Nuclei for C III and C IV which supply the eye muscles are located in the mesencephalon. Thickening of the *tectum* (roof) of the mesencephalon forms the paired *superior* and *inferior colliculi,* important elements in visual and auditory neural pathways. Large nerve tracts emerging from the base of the cerebrum as *cerebral peduncles* pass through the mesencephalon.

The *thalamus,* the principal structure of the diencephalon, is an elaborate integrating center receiving all neural impulses except those of olfaction. The hypothalamus initially presents optic cups (future eyes) and the stalk of the pituitary gland.

The telencephalon, highly complex in humans, comprises the remainder of the brain. A *longitudinal sulcus* (incomplete midline furrow) divides the telencephalon into *cerebral hemispheres.* Each consists of a *rhinencephalon* (in part a very primitive olfaction center) and *basal ganglia* (an elaborate high-order motor relay center). The basal ganglia and thalamus are just lateral to the third ventricle. The remainder of the telencephalon consists of the cerebral cortex (highly convoluted in the mature brain) and pathways of *association fibers* (interconnect various regions) and *commissural fibers* (connect opposite hemispheres). During an early period of rapid growth the telencephalon folds over on itself, thereby isolating a region called the *insula.* The covering folds are called the *opercula,* and in the fully developed brain, separation of the margins of the lateral sulcus reveals the insula. These margins contain areas vital to speech production and language integration. The cerebral cortex, formed by a very thin "bark," can be divided into functional regions such as somatic sensory, somatic motor, auditory language reception, bodily awareness, intelligence, and more. The elaborate cortex is what makes us human, and of course, so very wise.

FIGURE 7.2 DIFFERENTIATION OF THE NEURAL TUBE TO FORM THE BRAIN VESICLES AND SOME OF THEIR DERIVATIVES.

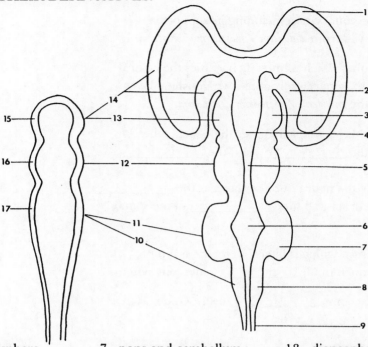

Identify:

1. cerebral hemisphere
2. basal ganglia
3. thalamus
4. third ventricle
5. cerebral aqueduct
6. fourth ventricle

7. pons and cerebellum
8. medulla oblongata
9. spinal cord
10. myelencephalon
11. metencephalon
12. mesencephalon

13. diencephalon
14. telencephalon
15. prosencephalon
16. mesencephalon
17. rhombencephalon

Quotes: "A burgeoning number of studies indicates that language learning begins long before infants utter their first words, probably within the womb upon hearing their mother's voice. . . . In some studies, for example, newborns listen longer to recorded speech in the mother's language than to utterances from another tongue."

Bruce Bower, 1997

"Fetal alcohol syndrome (FAS) is a distinct pattern of congenital physical and mental abnormalities. When first described in the 1970s, FAS was found in infants born to women who drank heavily during their pregnancies. Since then it has been identified even in offspring of moderate drinkers. Elements of this syndrome may include some or all of the following: low birth weight, poor coordination, facial deformities, heart defects, hyperactivity, and mental retardation. In the U.S. the full syndrome occurs in about one to three of every 1,000 live births. FAS ranks with Down syndrome and spina bifida as a major cause of mental retardation. Because the fetus metabolizes alcohol more slowly than the adult, the drug persists longer in the fetus than in the mother. The minimum amount of alcohol consumption required for causing FAS—or a less severe condition called fetal alcohol effect—is unknown; therefore, it is recommended that women abstain from alcohol throughout pregnancy and the preconception period."

Bruce D. Shepherd
1995 Medical and Health Annual

No. 7-5 EMBRYONIC DEVELOPMENT OF THE NERVOUS SYSTEM

1. The entire nervous system including its supportive
   cells, are derived from *endoderm / mesoderm / ectoderm*.

   1. _____

2. The neural tube, which ultimately becomes the central
   nervous system, separates from its own germinal layer
   and invades the *endoderm / mesoderm / ectoderm*.

   2. _____

   This germinal layer that surrounds the neural tube
   will become the _____ and the _____.

   _____

   _____

3. The hollow of the neural tube remains as the _____
   system of the brain and the _____ of the spinal
   cord.

   3. _____

   _____

4. The neural crest, which is lateral to the neural tube,
   gives rise to much of the _____ nervous system.

   4. _____

5. Neuroblasts become _____ cells. Glioblasts
   become _____ cells.

   5. _____

   _____

6. As cells in the neural tube develop, they migrate from
   the *inner to outer / outer to inner* layers of the
   tube, and are joined by axons of cells from the
   _____.

   6. _____

   _____

7. Glial cells, having migrated with the developing
   neurons, begin to surround the axons and dendrites,
   and to insulate some of them with _____.

   7. _____

8. Assignment of function (sensory or motor) is more
   clearly delineated in the *brain / spinal cord*.

   8. _____

9. Why does the central canal of the
   spinal cord become smaller?

   9. _____

   _____

10. The spinal cord is ready for reflex activity approxi-
    mately *2 / 4 / 6* months after conception.

    10. _____

11. A two-neuron reflex requires a/an _____
    and a/an _____ neuron.

    11. _____

    _____

12. A three-neuron reflex would have an additional
    "messenger," a/an _____ neuron.

    12. _____

No. 7-6 ADULT DERIVATIVES OF THE PRIMARY BRAIN VESICLES

DERIVATIVES	PRIMARY VESICLES AND SUBDIVISIONS					
	Prosencephalon		Mesencephalon	Rhombencephalon		Medulla Spinalis
	Telencephalon	Diencephalon	Mesencephalon	Metencephalon	Myelencephalon	Medulla Spinalis
pons				✓		
striate bodies (corpus striatum)	✓					
cerebral peduncles			✓			
thalamus		✓				
spinal cord						✓
cerebrum (cerebral hemispheres)	✓					
medulla oblongata					✓	
collicular structures			✓			
hypothalamus		✓				
rhinencephalon (structures in medial aspect of cerebral hemispheres)	✓					
cerebellum				✓		
CAVITIES						
central canal (major portion)						✓
third ventricle (major portion)		✓				
fourth ventricle (major portion)				✓		
cerebral aqueduct			✓			
lateral ventricle and part of third ventricle	✓					
fourth ventricle and part of central canal					✓	

Quote: "In the young embryo the brain and spinal cord make their appearance as a shallow groove, excavated down the midline of the back surface. The edges of this groove roll over and grow together, so that the whole of the cerebral axis assumes the form of a tube, which gradually zips itself shut from head to tail, leaving a microscopic pore at the lower end, which finally seals itself off. If the zipping instructions are not followed, the spinal cord fails to close, and the nervous system is left exposed at the lower end, forming the well-known condition of spina bifida."

Jonathan Miller, 1978

References and Sources

Asimov, I., *Understanding Physics*, Vol. 1. London: George Allen and Unwin Ltd., 1966.

Batshaw, M. L. and Y. V. Perret, *Children with Handicaps: A Medical Primer*, 2nd ed. Baltimore: Paul H. Brookes, 1986.

Berrill, N. J., *Biology in Action*. New York: Dodd Mead, 1966.

Bingham, B. J. G., M. Hawke, and P. Kwok, *Atlas of Clinical Otolaryngology*. St. Louis: Mosby–Year Book, 1992.

Bloom, F. E. and A. Lazerson, *Brain, Mind and Behavior*, 2nd ed. New York: W. H. Freeman, 1988.

Borden, G. J. and K. S. Harris, *Speech Science Primer: Physiology, Acoustics, and Perception of Speech*, 2nd ed. Baltimore: Williams and Wilkins, 1984.

Bower, B., "Everybody's Talkin'," *Science News*. 151, 1997, 276–277.

Carrell, J., and W. R. Tiffany, *Phonetics: Theory and Application to Speech Improvement*. New York: McGraw-Hill, 1960.

Chipps, E., N. Clanin, and V. Campbell, *Neurological Disorders*. St. Louis: Mosby–Year Book, 1992.

Crouch, J. E., *Functional Human Anatomy*, 3rd ed. Philadelphia: Lea and Febiger, 1978.

Daniloff, R., G. Schuckers, and L. Feth, *The Physiology of Speech and Hearing: An Introduction*. Englewood Cliffs, NJ: Prentice-Hall, 1980.

Darby, J. K. Jr. (ed.), *Speech Evaluation in Medicine*. New York: Grune and Stratton, 1981.

Darley, F. L., A. E. Aronson, and J. R. Brown, *Motor Speech Disorders*. Philadelphia: W. B. Saunders, 1975.

DeWeese, D. D., and W. H. Saunders, *Textbook of Otolaryngology*, 5th ed. St. Louis: C. V. Mosby, 1977.

Diamond, M. C., A. B. Scheibel, and L. M. Elson, *The Human Brain Coloring Book*. New York: Barnes and Noble Books, 1985.

Didio, L. J., A. *Synopsis of Anatomy*. St. Louis: C. V. Mosby, 1970.

Dorland's *Illustrated Medical Dictionary*, 28th ed. Philadelphia: W. B. Saunders, 1994.

Durant, J. D. and J. H. Lovrinic, *Bases of Hearing Science*, 2nd ed. Baltimore: Williams and Wilkins, 1984.

Fairbanks, G., *Voice and Articulation Drillbook*, 2nd ed. New York: Harper and Row, 1960.

Fant, G., *Acoustic Theory of Speech Production*. The Hague: Mouton, 1968.

Gerber, S. E., *Prevention: The Eitiology of Communicative Disorders in Children*. Englewood Cliffs, NJ: Prentice-Hall, 1990.

Gorlin, R. J., and J. J. Pindborg, *Syndromes of the Head and Neck*. New York: McGraw-Hill, 1964.

Gray, H., *Gray's Anatomy*, 36th & 38th British ed. (P. Williams and R. Warwick, eds.) Philadelphia: W.B. Saunders, 1980, 1995.

Guyton, A. C., *Textbook of Medical Physiology*, 4th & 6th ed. Philadelphia: W. B. Saunders, 1971, 1981.

Hewitt, P. G., *Conceptual Physics*, 6th ed. New York: Harper Collins, 1989.

Jacob, S. W., C. A. Francone, and W. J. Lossow, *Structure and Function in Man*, 4th ed. Philadelphia: W. B. Saunders, 1978.

Kent, R. D. and C. Read, *The Acoustic Analysis of Speech*. San Diego: Singular Publishing, 1992.

Kirchner, J. A. (ed.), *Vocal Fold Histopathology: A Symposium*. San Diego: College Hill Press, 1986.

Lass, N. J. (ed.), *Principles of Experimental Phonetics*. St. Louis: Mosby–Year Book, 1996.

Lass, N. J., L. V. McReynolds, J. L. Northern, and D. E. Yoder (eds.), *Speech, Language and Hearing* (3 vol.). Philadelphia: W. B. Saunders, 1982.

Liebman, M., *Neuroanatomy Made Easy and Understandable*. Baltimore: University Park Press, 1979.

Lim, D. J., C. D. Bluestone, J. O. Klein, and J. D. Nelson (eds.), *International Symposium on Recent Advances in Otitus Media With Effusion*, 3rd ed. St. Louis: C. V. Mosby, 1984.

Marieb, E. N., *Essentials of Human Anatomy and Physiology*. 4th ed. Redwood City, CA: Benjamin/Cummings, 1994.

Markel, H. and F. A. Oski, *The Practical Pediatrician*. New York: W. H. Freeman, 1996.

Marquardt, T. P., *Acquired Neurogenic Disorders*. Englewood Cliffs, NJ: Prentice-Hall, 1982.

McAleer, N., *The Body Almanac*. Garden City, NY: Doubleday, 1985.

McKeough, D. M., *The Coloring Review of Neuroscience*. Boston: Little Brown, 1982.

Miller, J., *The Body in Question*. New York: Random House, 1978.

Minifie, F., T. Hixon, and F. Williams (eds.), *Normal Aspects of Speech, Hearing, and Language*. Englewood Cliffs, NJ: Prentice-Hall, 1973.

Moore, K. L., *Clinically Oriented Anatomy*, 2nd ed. Baltimore: Williams and Wilkins, 1985.

Nelson, W. E., V. C. Vaughan, III, R. J. McKay, Jr., and R. E. Behrman (eds.), *Nelson: Textbook of Pediatrics*. Philadelphia: W. B. Saunders, 1979.

1995 *Medical and Health Annual*. Chicago: Encyclopedia Britannica, 1994.

Patten, B. M., *Human Embryology*. Philadelphia: Blakiston, 1946.

Pickett, J. M., *The Sounds of Speech Communication*. Baltimore: University Park Press, 1980.

Rahn, H., A. Otis, L. E. Chandwick, and W. Fenn, "The Pressure-Volume Diagram of the Thorax and Lung," *Amer. J. Physiol.*, 148, 1946, 161–178.

Restak, R. M., *The Brain*. New York: Bantam Books, 1984.

Rose, D. E. (ed.), *Audiological Assessment*, 2nd ed. Englewood Cliffs, NJ: Prentice-Hall, 1978.

Schopmeyer, B. B. and F. Lowe (eds.), *The Fragile X Child*. San Diego: Singular Publishing, 1992.

Shames, G. H., and E. H. Wiig (eds.), *Human Communication Disorders: An Introduction*. Columbus: Charles E. Merrill, 1982.

Sicher, H., and E. L. DuBrul, *Oral Anatomy*. St. Louis: C. V. Mosby, 1975.

Smith, A. *The Body*. New York: Viking, 1986.

Sundberg, J., "Acoustics of the Singing Voice," *Scientific American*, 236, March, 1977.

Titze, I. R., *Principles of Voice Production*. Englewood Cliffs, NJ: Prentice-Hall, 1994.

Tortora, G. J., *Principles of Human Anatomy*, 5th ed. New York: Harper Collins, 1989.

Travis, L. E. (ed.), *Handbook of Speech Pathology and Audiology*. New York: Appleton-Century-Crofts, 1971.

Van Nostrand's *Scientific Encyclopedia*. 7th ed. New York: Van Nostrand Reinhold, 1989.

Vennard, Wm., *Singing: The Mechanism and the Technic*, rev. ed. New York: Carl Fischer, 1967.

Webster's Third New International Dictionary, Springfield, MA: Merriam Webster, 1986.

Whitaker, L. A., and P. Randall (eds.), *Symposium on Reconstruction of Jaw Deformity*, Vol. 16. St. Louis: C. V. Mosby, 1978.

Wise, C. M., *Applied Phonetics*. Englewood Cliffs, NJ: Prentice-Hall, 1957.

Woodburne, R. T., *Essentials of Human Anatomy*, 5th ed. New York: Oxford University Press, 1973.

Yost, W. A., and D. W. Nielsen, *Fundamentals of Hearing: An Introduction*. New York: Holt, Rinehart and Winston, 1977.

Zemlin, W. R., and S. Stolpe, *The Structure of the Human Skull*. Champaign, IL: Stipes Publishing, 1967.

Answer Key

Chapter 1

1-1
1. superior
2. lateral
3. all proximal
4. both anterior
5. peripheral
6. central
7. dorsal
8. deep
9. posterior
10. rostral
11. ventral
12. caudal
13. lateral
14. all distal
15. inferior
16. superficial
17. superior

1-4
growth
reproduction
irritability
metabolism
spontaneous movement

1-5
1. lysosome
2. mitochondria
3. cell
4. nucleolus
5. nucleolus
6. cytology
7. cytoplasm
8. Golgi apparatus
9. tissue
10. DNA
11. centrosome
12. vacuole
13. protoplasm
14. endoplasmic reticulum
15. DNA
16. cell membrane
17. nucleus

1-7
1. single cell mucous gland which secretes mucin
2. flat, pavementlike

1-12
1. youngsters, it forms the growing skeleton
2. chondro

1-14
1. irregular
2. long
3. irregular
4. sesamoid
5. air-containing
6. long
7. short
8. sesamoid
9. flat
10. accessory
11. short

1-16
Elevations
1. crest
2. process
3. condyle
4. tubercle
5. head
6. tuberosity
7. spine
8. trochanter

Depressions
1. fissure
2. sinus
3. foramen
4. neck
5. fossa
6. sulcus
7. meatus
8. fovea

1-17
1. trabeculae
2. anastomose
3. lacunae
4. diaphysis
5. chondroblasts
6. bone marrow
7. perichondrium
8. canaliculi
9. aponeuroses
10. osteoblasts
11. matrix
12. lamellae
13. collagenous
14. diploe
15. interstitial growth
16. parenchyma
17. cortex
18. yellow marrow
19. periosteum
20. subcutaneous fascia
21. osteocyte
22. appositional growth
23. red marrow
24. epiphysis

1-19
1. edges of bone are like saw teeth
2. toothlike projections on opposing edges of bone
3. interlock on beveled edges

1-21
1. pivot
2. saddle
3. hinge
4. ball and socket
5. gliding
6. condyloid
7. pivot, ball and socket
8. condyloid and saddle

1-22
1. both
2. involuntary
3. striated

1-23
1. myofibril
2. fascia
3. myocardium
4. sarcoplasm
5. fusiform
6. perimysium
7. kinesiology
8. endomysium
9. sarcolemma
10. fasciculi
11. ephaptic conduction
12. epimysium
13. myoglobin

1-24
1. a
2. c
3. a
4. b, c
5. b, c

1-25
1. electrocardiogram
2. isotonic contraction
3. electromyography
4. rigor mortis
5. endoplasmic reticulum
6. isometric contraction
7. single muscle twitch
8. sarcomere
9. actin filament
10. myosin filament
11. muscle tone
12. ATP

1-28
1. gravity
2. antagonist
3. prime mover
4. lever arm
5. applied force
6. synergist
7. mechanical disadvantage
8. fixation muscle
9. mechanical advantage
10. fulcrum
11. resistance force

(1) applied force   (2) lever arm   (3) fulcrum

1-30 See descriptions in text.

1-32
1. location
2. location
3. function
4. geometric shape
5. general form
6. function
7. function
8. attachments
9. number of heads
10. attachments
11. location
12. number of heads
13. function
14. general form

1-33
1. long
2. irritability
3. electrochemical composition

1-34
1. latent period
2. motor unit
3. refractory period
4. axon
5. relaxation period
6. muscle end plate
7. motor unit
8. sarcoplasm
9. contraction period

(1) no   (2) yes   (3) yes   (4) yes   (5) yes

1-35
1. thromboplastin
2. erythrocytes
3. lymphocytes
4. blood
5. blood platelets
6. blood plasma
7. leukocytes
8. lymph

1-36
1. organ
2. parenchyma
3. yes
4. yes
5. lung
6. ear
7. larynx

1-37
1. skeletal—osteology (preservation and correction, orthopedics)
2. articular—arthrology
3. muscular—myology
4. digestive—splanchnology (stomach & intestines, gastroenterology)
5. vascular—angiology (heart, cardiology)
6. nervous—neurology
7. respiratory—pulmonology
8. urinary—urology
9. generative—gynecology reproductive
10. endocrine—endocrinology
11. integumentary—dermatology

1-38
1. cell
2. system
3. gland
4. organ
5. tissue
6. organ
7. cell
8. tissue
9. organ
10. tissue
11. cell
12. organ
13. gland, organ
14. organ*
15. tissue
16. cell
17. tissue
18. tissue
19. cell
20. tissue

*all of the skin and its appendages form a system

1-39
1. respiration   lungs, lower respiratory tract
2. phonation   vocal folds
3. articulation   lips, tongue, palate
4. resonation   pharyngeal, oral and nasal cavities

It ignores the important role of auditory and proprioceptive feedback. It does not adequately describe the synergy of speech production, but rather emphasizes an unrealistic temporal sequence of events.

Chapter 2

2-1
1. physical
2. mechanical
3. chemical

2-2
1. motion
2. gases
3. temperature
4. True
5. True

2-3
1. True
2. True
3. False
4. True
5. a. 4
   b. 2
   c. 1
6. a. False
   b. False
   c. True
   d. True
   e. True
7. True

2-4
1. zero
2. 1520 mm Hg, 11.2 liters, positive
3. x = 4, P = 3040 mm Hg, V = 5.6 liters, positive
4. negative, into, inhalation
5. positive, out of, exhalation
6. airflow

2-5
1. from top to bottom: nasal cavity, oral cavity, pharyngeal cavity, vocal folds, trachea, bronchi, lungs.
   Upper—above vocal folds
   Lower—below vocal folds
2. a. oral cavity
   b. trachea
   c. pharynx
   d. vocal folds
   e. larynx
   f. palate
3. vocal folds
4. warm (or cool in extreme heat), moisten, filter
5. tracheal
6. passive
7. thoracic (each lung is within a pleural cavity)
6. cough
7. thoracic

2-6
1. 4½
2. 1
3. anterior
4. posteriorly
5. hyaline
6. cricotracheal
7. mainstem bronchi
8. intratracheal
9. both, contraction
10. mucous
11. goblet
12. cilia, mucus
13. tracheotomy
14. tracheostomy

2-7
1. trachea, lung
2. bronchus
3. a. main stem
   b. secondary
   c. tertiary
4. trachea
5. bronchitis
6. is more in direct line with trachea and also somewhat larger in diameter

2-8
1. bronchioles
2. greater
3. terminal, alveolar, air sacs
4. muscular
5. alveoli, capillaries
6. by the large area of the capillary network and by the thin barrier which separates it from alveolar air

2-9
1. heart
2. properties of the alveoli
3. interface
4. collapse, surfactant
5. surface tension, surfactant
6. large
7. diaphragm
8. five
9. thoracic, lungs
10. a nonexistent, but potential space between the pleurae
11. provide friction-free surfaces which glide smoothly during breathing; form protective sacs so puncture will not collapse both lungs

2-10
1. surfactant
2. lung
3. pneumothorax
4. hilum
5. mediastinum
6. surface tension
7. pleurisy
8. pleurae
9. thorax
10. pulmonary alveolar epithelium
11. pulmonary ligament
12. pericardium
13. pleural sinuses
14. in situ

2-11
1. increases
2. a. liquid
   b. permeable
   c. visceral
3. manometer
4. lubricate
5. is exceeded by, increase
6. below

2-12
1. vertebral column, rib cage, pelvis
2. True

2-13
1. sacrum
2. coccyx, tailbone
3. spinal cord
4. 7th
5. articulations, ligaments
6. a. thoracic
   b. cervical
   c. lumbar

2-14
1. cervical
2. spina bifida
3. vestigial
4. lumbar
5. scoliosis
6. coccygeal
7. rudimentary
8. axis
9. dens, odontoid
10. corpus
11. atlas
12. lordosis
13. pedicles
14. kyphosis
15. sacrum
16. vertebra
17. thoracic
18. vertebral foramen

2-15
1. a. 12 thoracic vertebrae
   b. sternum
   c. 12 pairs of ribs
2. vertebrae
3. inferior
4. horizontal
5. gliding
6. one
7. two

2-16
1. inhalation
2. diaphragm
3. raised
4. sternum
5. anteroposterior, lateral
6. expiratory, inspiratory

2-17
1. coxal, hip
2. coxal, sacrum, coccyx
3. acetabulum
4. gluteus maximus
5. inguinal
6. ilium, ischium, pubis
7. ischium
8. ilium
9. ilium
10. Most of the muscles of the abdominal wall attach to the ilium. The pelvis functions as a supporting basin for the abdominal viscera.

2-18
1. clavicle, scapula
2. upper limbs, arms
3. scapula
4. clavicle
5. acromion
6. jugular, suprasternal

2-19
1. xiphoid process, ensiform process
2. coxal bone
3. sternal angle
4. coracoid
5. pectoral
6. costal
7. vertebrosternal ribs
8. vertebral ribs
9. manubrium
10. ischium
11. acetabulum
12. vertebrochondral ribs
13. clavicle
14. jugular notch, suprasternal notch
15. sternum
16. scapula
17. acromion
18. corpus
19. omo-
20. inguinal ligament

2-20
1. a. three
   b. pleural
   c. negative
   d. atmospheric
   e. positive
2. passive, nonmuscular
3. five
4. .5–.75
5. inspiration
6. abdominal

2-21
1. thorax, abdomen
2. central tendon
3. diaphragm
4. lungs and heart, both
5. liver
6. unpaired
7. bilateral

2-24
1. aponeurosis
2. potential energy
3. pericardium
4. crura
5. foramen, hiatus
6. diaphragm
7. central tendon
8. radiography
9. levatores costarum
10. intercostal
11. palpation
12. forced exhalation
13. EMG
14. breves
15. kinetic energy

2-25
1. a. forward
   b. vertically
2. a. increase
   b. decrease
   c. decrease
   d. increase
3. ½, 4
4. inhalation, exhalation
5. True
6. False
7. True
8. False
9. during expulsive efforts requiring fixation of the thoracic-abdominal system, particularly when transmission of abdominal pressure to the thoracic cavity is needed as in coughing, sneezing, laughing

2-26
1. a. thoracic
   b. ribs
   c. ribs
   d. postural
   e. speech production
2. a. disadvantage
   b. lost
   c. gained
   d. common
3. lower, elevate, inhalation
4. upper, depress, exhalation
5. decreased
6. both
7. expiratory
8. decreases
9. is
10. expiratory
11. False EMG can determine muscle activity, but that does not necessarily prove the function of a muscle.

2-28
1. a. temporal
   b. opposite, flex
   c. sternum, clavicle, inhalation
2. a. outer
   b. cervical
   c. same, flex
   d. elevate, inhalation

2-30
1. linea alba
2. abdomen
3. lumbodorsal fascia
4. inguinal ligament
5. abdominal aponeurosis
6. linea semilunaris

2-32
1. posterior
2. exhalation
3. inhalation
4. lumbar

2-35
1. pulmonary subdivisions
2. breath group
3. yawn
4. spirogram
5. dead air
6. spirometer
7. torque
8. residual air
9. dead air space
10. pulmonary compliance

2-36

Characteristics of spirograms when lung compliance is	Excessive (e.g., emphysema)	Lessened (e.g., silicosis)
Expiratory Reserve Volume	elevated	decreased
Functional Residual Capacity	elevated	decreased
Vital Capacity	decreased	decreased
Total Lung Capacity	elevated	decreased
Tidal Volume	unchanged	decreased
Inspiratory Reserve Volume	decreased	decreased
Inspiratory muscle activity	heightened	heightened
Expiratory muscle activity during quiet breathing	required	unchanged
Breathing rate		may increase

Pulmonary Measures Effects of:	Pressurized Environment	Partial Vacuum
Tidal Volume	decreased	unchanged
Inspiratory Reserve Volume	decreased	decreased
Expiratory Reserve Volume	unchanged (theoretically)	decreased
Vital Capacity	decreased	decreased
Total Lung Capacity	decreased	unchanged

2-37
1. 500, 12, liters
2. increases, increases, oxygen expenditure increases as work increases.
3. oxygen, carbon dioxide
4. yes, helps prevent wide fluctuations of concentrations of oxygen and carbon dioxide in the blood stream
5. decrease, abdominal viscera press diaphragm upward; increase in pulmonary blood volume decreases pulmonary air space
6. height/weight, respiratory muscle strength, pulmonary compliance
7. increases, decreases
8. a. larger d. lower
   b. greater e. much higher
   c. greater
9. a. decreases c. decreases
   b. decreases d. increases

2-38
1. outward 2. inward
3. downward, The lungs are closely bound to the diaphragm.
4. 0 8. a. enlarge
5. abdominal b. collapse
6. exhalation c. descend
7. inhalation d. distend

2-39
1. inhalation, exhalation
2. exhalation, inhalation
3. inhalation, exhalation
4. shorter
5. air friction

2-40
1. airway resistance, muscular pressure, relaxation pressure
2. relaxation
3. zero
4. positive, passive, atmospheric
5. low
6. lung, alveolar
7. linear, nonlinear, abruptly
8. lungs, chest wall (thorax)
9. chest wall, lungs
10. True
11. a. active
    b. passive
12. a. passive
    b. active
13. muscular, relaxation
14. pressure-volume
15. hiccup, sneeze
16. Inspiratory and expiratory forces can be generated simultaneously and may yield pressures which might mistakenly be accounted for by relaxation pressure.

2-41
1. low
2. abs vocal folds, sometimes lips
3. low
4. low

2-41
cont'd
5. low
6. low
7. subs — vibrating vocal folds
8. abs — elevated palate, compressed lips, then vocal fold vibration
9. abs — elevated palate, compressed lips
10. subs — elevated palate, alveolar ridge - tongue
11. subs — elevated palate, alveolar ridge - tongue, vocal fold vibration

(1) voltage, resistance
(2) alveolar pressure, airway resistance
(3) increase

2-42
1. tracheal puncture, passing a balloon into the esophagus

2-42
cont'd
2. relatively constant
3. high
4. checking action
5. increases
6. decreases, inverse
7. True
8. linear, nonlinear
9. singing
10. loud

2-43
1. pressure
2. a. True
   b. False
   c. False

2-44
1. True
2. intercostals
3. excessive
4. insufficient
5. quiet breathing
6. both speech and singing

Chapter 3

3-1
1. larynx, vocal folds
2. a. trachea   c. 3–6
   b. hyoid bone   d. anterior
3. thyroid

3-2
1. respiratory
2. prevents escape of air from lungs, prevents foreign substances from entering larynx, forcefully expels foreign substances threatening to enter larynx or trachea
3. thoracic   4. yes

3-3
1. False
2. wide
3. glottal, laryngeal, vocal fold
4. see text
5. a. 125
   b. 210
6. contracts, releases

3-4
1. axial
2. tongue
3. tongue
4. extrinsic
5. not directly attached to another bone
6. sling, strap
7. third
8. corpus
9. greater horns (cornua)
10. lesser horns (cornua)

3-6
1. raphe
2. commissure
3. symphysis
4. triticial
5. glottis spuria
6. laryngeal prominence, thyroid prominence
7. digastric
8. cricoid
9. synergy
10. glottal stop, glottal arrest
11. mental, genio-
12. glottal chink, glottis, rima glottidis

3-6
cont'd
13. mandible
14. atavism
15. hyoid
16. stylo-
17. cuneiform
18. mylo-
19. thyroid
20. glottal attack, glottal stroke
21. arytenoid
22. pars
23. aditus
24. deglutition
25. glottis
26. cornua
27. colliculus
28. omo-

3-7
1. angle
2. thyroid or laryngeal prominence
3. 90
4. longer
5. hyoid
6. cricoid

3-8
1. lower
2. first, uppermost; cricotracheal
3. band, arch, signet, lamina
4. pivot, either

3-9
1. pyramid
2. glottis (opening between the folds)
3. cricoid
4. vocal, anterior, base
5. muscular
6. corniculate

3-10
1. hyoid, root
2. thyroid, hyoid
3. children
4. tongue, epiglottis
5. valleculae
6. fat pad
7. no
8. no
9. larynx
10. pitch

3-11
1. aryepiglottic, larynx
2. cuneiform
3. cuneiform
4. vestigial
5. aryepiglottic, larynx

3-12 1. two, all internal adjustments of vocal folds mediated through them
2. lamina, convex   6. yes
3. muscular, concave   7. ligaments
4. abducted, adducted
5. saddle (diarthrodial)

3-13 1. arch, inferior
2. rotational, gliding, increase tension, thus increasing pitch
3. ceratocricoid or capsular

3-14 1. extrinsic   2. intrinsic
ELM 1. hyothyroid, hyothyroid, triticial
2. hyoid bone, epiglottis
3. cricotracheal
ILM 1. elastic   2. conus elasticus
3. quadrangular, aryepiglottic
MM 1. mouth and pharynx, trachea
2. vocal folds

3-15 1. aditus
2. cricoid
4. sinus
5. supraglottal, subglottal
6. ventricular (false) folds, aditus laryngis
7. true folds, ventricular folds
8. thyroid, arytenoid   9. false, wider
10. subglottal, pharynx, help remove mucus and foreign matter from the respiratory tract

3-16 2. variable opening between the vocal folds
3. thyroarytenoid, vocal, elasticus
4. membranous, intermembranous
5. cartilaginous, intercartilaginous
6. anterior, membranous
7. less

3-17 1. hyoid, suprahyoid, infrahyoid

3-19 1. thyrohyoid   6. digastric
2. hyoglossus   7. sternothyroid
3. mylohyoid   8. geniohyoid
4. genioglossus   9. stylohyoid
5. sternohyoid   10. omohyoid
11. inferior pharyngeal constrictor

3-20 1. compression, tension   2. close

3-21 1. posterior cricoarytenoid
2. oblique arytenoid   4. thyroarytenoid
3. lateral cricoarytenoid 5. cricothyroid
6. transverse arytenoid

3-22 1. location of the larynx, illumination, rapidity of vocal fold vibration
2. direct, indirect   4. stroboscope
3. nose, trachea   5. high-speed
6. transillumination
7. Hyaline cartilage does not absorb x-rays well, and in children the cartilage has not begun to ossify.
8. laminagraphy, sectional radiography, both
9. laminagraphy, stroboscopy, slow-motion
10. hook-wire, to avoid contaminating effects of adjacent muscles
11. Because EMG indicates only that a muscle is active, you must have an understanding of anatomy to determine probable consequences of muscle activity.
12. air flow
13. tracheal puncture, esophageal balloon, tracheal puncture (because it is a direct measure and is not affected by pleural surface pressure)

3-23 1. 100 Hz 2. 400 Hz 3. 50 Hz

3-24 1. prephonation, attack
2. prephonation, increases, increases
3. lateral cricoarytenoid, arytenoid, posterior cricoarytenoid
4. directly   8. increase, decrease
5. attack   9. negative, together
6. approximated   10. before
7. Bernoulli   11. no
12. decrease in subglottal pressure, vocal fold tissue elasticity, Bernoulli effect

3-26 1. opening (longest), closing, closed (shortest)
2. area, time   4. speed
3. synchrostroboscopy 5. open
6. speed because sometimes the glottis never closes; speed quotient gives more information about vibratory characteristics
7. posterior
8. horizontal, vertical
9. first, first, vertical

3-27 1. two
2. fundamental frequency, male 130, female 220
3. pitch level
4. negatively
5. They cannot be sustained.
6. natural, optimum
7. habitual   8. ¼

3-28
1. longer
2. a. increases
   b. decreases
3. a. decrease
   b. decrease
4. no
5. tension
6. cricothyroid, thyroarytenoid, posterior cricoarytenoid
7. antagonists, arytenoid
8. breathy, Folds may not approximate at vocal processes
9. low
10. resistance, impedance
11. insignificant
12. habitual

3-29
1. lower
2. mass, tension
3. elasticity, thyroarytenoid
4. lateral cricoarytenoid
5. at the high and low extremes of the pitch range
6. algebraic, vector

3-30

```
              →TRA←
              →OBA←
  ↑PCA                    PCA↑

  ↓LCA                    LCA↓
       →LCA      LCA←
       ←PCA      PCA→

       ↑THA      THA↑
       ↓CTH      CTH↓
```

3-32
1. 70, 30
2. the rate of air-flow, glottal resistance is already near its maximum at high pitch levels
3. true, at high intensity levels the folds remain in the closed phase considerably longer, thus leaving less time for air flow to occur
4. a. subglottal pressure   b. glottal resistance

3-33
1. subglottal, supraglottal
2. glottal, supraglottal
3. a. increase
   b. decrease (approach zero)
   c. expand
4. decreases, inverse

3-34
1. vibrato
2. register
3. pitch
4. falsetto, loft
5. tremolo
6. modal pitch range
7. damping
8. fundamental frequency
9. laryngeal whistle
10. trill
11. glottal fry, pulse register

3-35
1. yes
2. The sounds produced by a lower voice have lower fundamental frequencies and thus more harmonically related overtones.

3-36
1. 2
2. habitual
3. air, 15-25
4. sound pressure level, 50
5. jitter
   a. rough
   b. male
6. noise
7. resonance or articulation

3-37
1. No. The folds do not appear to vibrate to any great extent and do not approximate with any force.

3-38
1. year
2. In males the vocal folds usually grow much more rapidly and also thicken.
3. no, the posterior attachment of the folds is not fixed.
4. lower, rise
5. decrease, deterioration of muscle tissue, increase of connective tissue in folds, and ossification of thyroid and cricoid cartilage.

3-39
A. 4   D. 6   G. 6   J. 1
B. 1   E. 2   H. 4   K. 2
C. 5   F. 3   I. 3   L. 5

3-40
1. neurochronaxic
2. myoelastic-aerodynamic
3. yes
4. mucoviscoelastic-aerodynamic
5. mechanical
6. synthesis
7. mathematical
Specified properties or behaviors of the larynx may be systematically varied and evaluated very precisely with greater ease, with few constraints, and in far less time.
8. oscillator, midline   9. vertical
10. The mucoviscoelastic-aerodynamic theory emphasizes the effect of the relatively loose coupling between the vocal ligament and the mucous membrane which lines the larynx.
11. mucous, vocalis, vocal
12. articulation

4-1
1. tone (buzz)
2. partial
3. formants
4. True

4-2
1. vertex
2. meatus
3. cranium, facial skeleton
4. occiput
5. lacri-
6. turbinate
7. cranium
8. frons
9. temple
10. calvaria
11. orbit
12. septum

4-3
1. a. semilunar
   b. dental alveoli
   c. pterygoid
   d. ramus
   e. corpus
   f. condyle
   g. mental
   h. lingula
   i. coronoid
   j. symphysis
   k. concha
   l. mylo-
2. mental
3. alveolar
4. posterior
5. It houses the teeth and provides attachment for muscles of the tongue and other structures. The movement of the mandible and its contained tongue modifies the size and acoustic characteristics of the oral cavity.

4-4
1. a. upper
   b. nasal
   c. orbital
2. roof
3. pyramid
4. frontal, zygomatic, alveolar, palatine
5. The bones are hollow and contain extensive maxillary sinuses
6. incisive
7. premaxilla
8. mandible, vomer

4-5
1. bridge
2. palatine
3. palatine
4. lacrimal
5. zygomatic
6. nasal conchae, turbinate
7. vomer bone, cartilaginous

4-6
1. coronal
2. styloid
3. zygo-
4. raphe
5. hypophysis
6. temporal
7. pterygoid
8. crista galli
9. parietal
10. glabella
11. sella
12. antrum
13. ethmoid, cribriform
14. mastication
15. clinoid
16. rostrum
17. glosso-
18. tympanic
19. sphenoid
20. lambdoid
21. lamellar

4-6 *cont'd*
22. squama
23. labyrinth
24. petrous
25. cecum
26. mastoid
27. nuchal
28. hamulus
29. chiasma

4-7
Ethmoid
1. nasal, cranial
2. olfactory, nasal
3. conchae

Frontal
1. forehead
2. orbital, nasal
3. ethmoid

Parietal
1. cranium

Occipital
1. cranium
2. foramen magnum
3. condyles

Temporal
1. cranium, braincase
2. zygomatic
3. petrous
4. air
5. petrous
6. styloid
7. by a freely movable joint (temporomandibular joint)

Sphenoid
1. one body, two greater wings, two lesser wings, two pterygoid processes
2. ethmoid, occipital
3. body
4. ethmoid
5. pituitary gland
6. greater wing
7. pterygoid process
8. provide attachments for important muscles and ligaments
9. the body of the sphenoid forms the geographic center of the skull

4-9
1. Little or no effect, other than possible minimal contribution to resonance
2. no
3. tympanic

4-10
1. buccal
2. oral
3. oral, pharyngeal
4. esophagus
5. tympanic, oral, laryngeal, nasal
6. oropharynx, nasopharynx
7. oropharynx, laryngopharynx
8. skull, 6th, cricoid

4-11
1. nasalis, depressor septi
2. bridge
3. maxillae, palatine bones
4. root
5. apex
6. nares
7. alar
8. respiratory
9. olfactory

4-11
cont'd
10. nasal dilators, quadratus labii superior (angular head)
11. procerus   12. choanae
13. nasal turbinates (conchae), nasal meatuses
14. ciliary action   15. septal

4-12
1. tongue tip
2. thin outer tissue reveals the hue of the vascular tissue
3. lingual   7. infants, suckling
4. alveolar, mandible   8. saliva
5. alveolar, maxilla   9. vowels
6. lips   10. consonants

4-15
1. growth, calcification, eruption, attrition
2. osteoclasts, osteocytes, osteoblasts
3. never during life   7. bicuspids,
4. as soon as it erupts   premolars
5. two   8. molars
6. second

4-16
1. maxillary   3. open, closed
2. diastema

4-17
1. taste, mastication (chewing), deglutition (swallowing)
2. complex arrangement of muscles and high innervation of muscle fibers
3. root, blade
4. human beings are upright animals, adds greater versatility
5. papillae, mucous, lymph   7. corium
        8. no
6. lingual

4-20
1. rapidity   3. result or cause
2. tongue-tip slightly faster   4. restrict
        5. subluxation
6. movement of the left and right joints is necessarily simultaneous

4-21
1. digastric ✔, geniohyoid ✔, mylohyoid ✔, lateral pterygoid
2. masseter, medial pterygoid, temporalis
3. masseter, temporalis   8. digastric
        9. lateral pterygoid
4. masseter, medial pterygoid   10. masseter
        11. geniohyoid
5. sternohyoid   12. medial pterygoid
6. mylohyoid   13. lateral pterygoid
7. temporalis   14. masseter

4-22
1. It modifies the degree of coupling between the nasopharynx and the rest of the vocal tract.

4-22
cont'd
2. a bony outgrowth along the site of the intermaxillary suture found in approximately 20% of population.
3. The width of the maxillary arch may give you an important clue. A person with a very narrow maxillary arch is more likely to have a highly vaulted palate than is a person who has a wide maxillary arch.
4. thinnest   7. nasals (m) (n) (ŋ)
5. velum   8. lowered
6. palatal aponeurosis

4-24
1. usually very little, but in some cases may affect nasal resonance and/or move tongue forward

4-25
1. respiratory, digestive 4. resonator
2. 4¾, 1⅗   5. no
3. circular, sphincterlike

4-26
1. inferior constrictor 11. mylopharyngeus
2. superior constrictor 12. chrondropharyn-
3. middle constrictor   geus
4. pharyngeal aponeurosis   13. pterygopharyngeal
        14. palatopharyngeus
5. cricopharyngeus   15. stylopharyngeus
6. glossopharyngeus   16. salpingopharyn-
7. buccopharyngeus   geus
8. stylopharyngeus   17. cricopharyngeus
9. ceratopharyngeus   18. pharynx
10. thyropharyngeus

4-27
1. oral, nasal   2. high
3. adjacent to nonnasal consonants
4. no, velopharyngeal closure usually occurs above the level of Passavant's pad
5. greater compensatory pharyngeal movement
6. medial movement of lateral walls
7. sphincterlike

4-28
1. deglutition   2. dysphagia

4-29
1. measurement of skull structures, vital staining of bones of living animals, serial x-ray studies
2. infancy
3. more
4. downward, forward
5. X=2. 66.6%, higher, may indicate inadequate tissue for velopharyngeal closure; removal of adenoids may result in irreversible hypernasality
6. greater development of alveolar bone

4-29
cont'd
7. an unossified membranous region in an infant skull, a "soft spot" 8. after 2½ yrs. Would restrict growth of the brain, probably resulting in mental retardation.
9. they are growth sites of the maxillary complex
10. length
11. incisors, molars
12. in infancy
13. both

4-30 about 77.6%. See norms text page 286.

4-31
1. a. air-pressure sensors
   b. pneumotachograph
   c. electromyography
   d. spirometer
2. radiation hazard (exposure)
3. filming time, radiation exposure
4. indirect palatography
5. a. ultrasound   b. strain gauge

4-32
1. laryngeal buzz, laryngeal tone
2. partial
3. frequency
4. volume velocity
5. overtone
6. damping
7. stop
8. voiced sound
9. articulation
10. harmonic
11. natural frequency
12. resonance
13. complex tone
14. fundamental frequency
15. tone
16. forced vibration
17. formants
18. fricative
19. spectrum
20. period
21. unvoiced sound
22. formant bands

4-33
1. no, the vocal folds cannot be isolated from the vocal tract
2. air column
3. a. short
   b. vocal fold vibration
   c. highly
   d. vocal folds, lips (mouth opening)
4. frequency (pitch), intensity (loudness), duration (length)
5. resonates
6. size, configuration (shape)

4-34
1. decreases   2. inverse, increases
3. a. 1000, 3000, 5000 Hz
   b. 250, 750, 1750 Hz
4. a. 80 cm, 425, 1275, 2125
   b. 40 cm, 850, 2550, 4250
   c. 160 cm, 212.5, 637.5, 1062.5
5. formant, maximal
6. men usually have longer vocal tracts
7. high, low

4-35
1. provide a standard reference for vowels used in all languages and describe the physiologic limits of tongue position for vowels
2. [u] who'd   [i] heed
   [o] hoed   [e] hayed
   [ɔ] hawed   [ɛ] head
   [ɑ] hod   [a] hard (New England dialect)
3. See Fig. 4-125, text page 301.
4. tense   5. diphthong

4-36
1. neutral
2. A formant is a physical property of the vocal tract; A spectral peak is a graphic representation of a formant
3. no   4. no   5. yes
6. length, location of constriction, degree of constriction
7. inversely   9. tongue, lips
8. lip rounding/protrusion, depression of larynx

4-38
1. uvular, pharyngeal
2. cognates, e.g., p/b, s/z, t/d, k/g
3. See Table 4-6, text page 305.
4. (a) 4 4 u 1 e 2 a 6 5 i o 4
   (b) 4 u 1 5 e 6 4 i 2 e (d) 6 i 4 e 4 3 e 5 i a
   (c) 1 i o 1 e 3 i 4 i 4 e (e) 6 o 4 o 4 1 o u 3
                              (f) 1 i o 1 4 y

4-39
1. nasals
2. affricates
3. stops
4. liquids
5. fricatives
6. glides
7. continuants
8. obstruents

4-40
1. burst release, vocal fold vibration
2. aspirated—audible friction after the release of the sound; unaspirated— glottis closes, no breath expelled
3. in [m] and [n] the oral and nasal cavities, operating as parallel resonant systems increase length of acoustic tube; oral cavity not a resonant system in [ŋ]
4. both, resonances

4-41
1. auditory
2. phonemes
3. indistinct
4. suprasegmental
5. transitions
6. burst
7. coarticulation
8. True
9. children, once speech is well established, kinesthetic feedback can maintain adequate articulation
10. comparator, compensatory

5-1
1. endocrine system
2. nervous system
3. central nervous system
4. peripheral nervous system
5. brain
6. neuron
7. neurotransmitter
8. axon
9. dendrite
10. afferent
11. efferent
12. nerve
13. nerve tract
14. synaptic cleft

5-3
1. medulla oblongata
2. cerebellum
3. pons
4. cortex
5. thalamus
6. cerebral aqueduct
7. hypothalamus
8. spinal cord
9. cerebrospinal fluid
10. cerebrum
11. gyri
12. sulci
13. lobes
14. basal ganglia
15. brain stem
16. meninges
17. dura mater
18. arachnoid mater
19. pia mater

5-4
1. outside the bony confines of the skull and spinal column
2. cranial and spinal nerves, autonomic nervous system
3. sensory, motor, mixed
4. mixed
5. brain stem
6. outside
7. gray, spinal cord
8. outside
9. termination
10. structure, function, distribution
11. autonomic nervous system
12. cranial and spinal nerves

5-5
1. visceral efferent system

5-6
1. meninges, cerebrospinal fluid, lubricate, moisten and protect the brain and spinal cord
2. venous, venous blood, cerebrospinal fluid
3. leptomeninges
4. cisterns
5. granulations (villi)
6. trabeculae, arachnoid
7. because it is highly vascular
8. choroid, ventricles

5-8
1. a. choroid
   b. ventricles
   c. subarachnoid
   d. brain
   e. arachnoid, sagittal
2. absorbed, drained
3. headache, slowed pulse and respiratory rate, loss of consciousness, in very young children chronic elevation results in hydrocephalous

5-9
1. gray
2. claustrum
3. radiata, internal, cerebri
4. cerebral cortex, thalamus, dopamine
5. a. increased
   b. resting
   c. involunatry
   d. athetosis
   e. Parkinson's
   f. diminution

5-10
1. cerebral hemispheres (cerebrum)
2. thalamus, ventricle, pulvinar
3. auditory
4. gray, nuclei
5. white
6. receives all neural impulses (except for olfactory sense), directly or indirectly from all parts of the body. It also receives impulses from the cerebellum, cerebral cortex, and many adjacent nuclei.
7. capsule, cerebral cortex

5-11
1. pineal body
2. pituitary gland
3. optic chiasma
4. neurohypophysis
5. posterior commissure
6. mammillary bodies
7. pineal body

5-12
1. forebrain, hindbrain
2. stem

5-13
1. a. stem
   b. spinal cord
   c. pons
2. descending, motor, efferent
3. spinal cord
4. crossed
5. direct
6. cerebellum
7. ascending, sensory, locomotion
8. cerebellum, cerebellar
9. reticular, cranial
10. respiration, circulation

5-14
1. stem
2. medulla oblongata, anteriorly
3. peduncles, cerebellum
4. medulla oblongata, cerebral
5. medulla oblongata, reticular
6. cerebellum

5-15
1. hind
2. oval, constriction
3. cerebelli
4. arbor
5. vermis, hemispheres
6. cerebrum

5-16
1. outside
2. voluntary and involuntary
3. cerebral, feedback
4. right, ipsilateral
5. dysfunction of input may affect cerebellar output
6. high, cortex

5-17
1. central canal
2. cauda equina
3. filum terminale
4. funiculi
5. nuclei, laminae, horns
6. intervertebral foramen

5-18
1. cervical, lumbar
2. right (90°), obtuse
3. extremities
4. 31–32
5. sulcus (fissure)
6. H, butterfly
7. diagnostic

5-19
1. a. pyramidal, granular
   b. motor
   c. association
2. surgical eradication of tissue, study of disease processes, electrical stimulation of cortex, comparative physiology
3. cortical
4. a. inverse
   b. direct
   4. c. homunculus
5. a. contralateral
   b. medially
   5. c. pyramidal, extrapyramidal
      d. Broca's
6. a. similar
   b. posterior
   c. temporal
   d. Wernicke's
   6. e. arcuate
      f. vision
      g. association
      h. nausea, vertigo
7. suppressor
8. a. left
   b. Wernicke's
   c. Broca's
   8. d. Wernicke's, Broca's
      e. children

5-21
1. a. reticular
   b. thalamus, cerebral cortex
   c. consciousness
2. a. diminishes
   b. REM (rapid eye movement)
3. large

5-22
1. within
2. outside

5-25
1. gray
2. sensory, peripheral
3. motor, central
4. spinal
5. spinal
6. dermatome, segmental
7. intervertebral, rami, sensory and motor
8. posterior
9. plexus, cranial
10. phrenic
11. brachial, chest, upper limb
12. lumbosacral

5-26
1. smooth, cardiac
2. involuntary
3. two, one
4. both
5. sympathetic
6. parasympathetic
7. The sympathetic preganglionic fibers arise in the spinal cord and synapse in the trunk

5-26 *cont'd*
ganglia on either side of the vertebral column. The parasympathetic preganglionic fibers arise in the brain and sacral region of the spinal cord and continue to the ganglion cells near the structures they supply.
8. a. tectal
   b. bulbar
   c. sacral
9. facial, glossopharyngeal, vagus

5-27
1. a. processes
   b. away from
   c. muscle, gland
   d. axon, efferent, never
   e. termination
   1. f. brushes, knobs or boutons
      g. neurotransmitter
      h. neurofilaments
      i. action
2. toward
3. a. peripheral
   b. all
   2. c. regeneration
4. Ranvier
   5. never
6. a. sometimes
   b. distally, Wallerian
   c. scar
   6. d. retrograde
      e. microglial (phagocytic)
      f. scar neuroma
7. a. sheath (tunnel)
   b. 12-18, atrophied

5-28
1. all
2. neurons, sensory receptors, muscle cells
3. the same
4. different
5. balanced
6. ion
7. a. + −
   b. + +, − −
   c. + −
   d. + −
8. a. number of
   b. distance between
   c. volts
9. a. conductor
   b. current
   c. amperes
   d. ohms
10. a. volts
    b. membrane, outside
    c. permeability
11. a. sodium
    b. potassium and chloride
12. a. membrane
    b. sodium
    c. potassium
    d. resting

5-29
1. electricity
2. a. depolarization
   b. decreases
   c. reversed
   d. recovery
   e. sodium-potassium pump
3. a. rheobase
   b. excitation, chronaxie
   c. excitability
4. a. after
   b. spike
   c. action
5. a. False
   b. True
   c. True, answer is (a)
6. tetany

5-29   7. a. direct   c. Ranvier
cont'd   b. myelinated   d. unmyelinated
             e. myelinated, outward current flows only
                 at nodes; jumps from node to node.

5-30   1. synapses, no   5. oxygen
         2. synaptic   6. temporal
         3. 100,000   7. spatial
         4. a. neurotransmitter   8. contraction
             b. terminal   9. inhibition, central
             c. postsynaptic
             d. enzyme

5-34   6, 1, 5, 3, 2, 4

5-35   1. proprioception   2. touch

5-36   1. voluntary,   7. voluntary
             corticospinal   8. fine
         2. limbs, cranial   9. poorly
         3. cerebral   coordinated,
         4. crossed   weaker, slower
         5. direct   10. spasticity
         6. above

5-37   1. pyramidal   3. direct, indirect
         2. coordinating   4. excite or inhibit

5-38   1. both   2. brain stem
         3. carbon dioxide in the blood, response of
            chemoreceptors to chemical composition of
            the blood, activity of stretch receptors in the
            lungs

5-38   4. Hering-Breuer,   6. inspiration
cont'd   pneumotaxic   7. inspiratory,
         5. blood   expiratory

5-39
Tongue
1. hypoglossal, bilateral
2. accessory, palate
3. important in the execution of finely
    coordinated articulatory movements
4. facial, glossopharyngeal, vagus

Mastication
1. trigeminal   3. bilateral
2. trigeminal, hypoglossal

Pharynx
1. vagus   2. glossopharyngeal

Palate
1. mandibular   3. maxillary, facial
2. accessory

Larynx
1. vagus
2. trigeminal, facial, hypoglossal

5-40   1. endocrine system   11. pituitary gland
         2. hormones   12. thyroxin
         3. pancreas   13. pituitary gland
         4. adrenal glands   14. parathyroid
         5. thyroid gland   glands
         6. pituitary gland   15. adrenal glands
         7. parathyroid glands 16. gonads
         8. thyroxin   17. pancreas
         9. gonads   18. thyroxin
        10. pancreas   19. adrenal glands

Chapter 6

6-1   1. a. large   6. fluid
            b. small   7. mass
            c. small   8. compressed,
            d. large   rarefied
         2. simple harmonic   9. longitudinal
         3. none   10. impart energy
         4. 1/60 second   11. displacement
         5. equilibrium, constant

6-2   1. A, C, E   4. B, D
         2. A, E   5. + y
         3. C

         6. $\dfrac{1}{.1}$ = 10 Hz   7. $\dfrac{1}{.002}$ = 500 Hz

         8. $\dfrac{1}{2000}$ = .0005 sec. (50 milliseconds)

6-2   9. $\dfrac{1}{10,000}$ = .0001 sec. (10 milliseconds)
cont'd

6-3   1. displacement   13. amplitude
         2. resonance   14. Hooke's Law
         3. mass   15. maintained
         4. period   vibration
         5. forced vibration   16. cancellation
         6. damping   17. rarefaction
         7. vibratory motion   18. critical damping
         8. compression   19. frequency
         9. inertia   20. phase
       10. simple harmonic   21. wave motion
            motion, sinusoidal 22. root-mean-square
            motion   amplitude
       11. Brownian movement 23. vibratory motion
       12. free vibration   24. periodic

6-4 1. a. parallel   b. a bit later

6-5 1. C   5. C   9. B
2. B   6. A   10. B
3. A   7. A   11. C
4. B   8. C

6-6 1. λ 3. 1130
2. 360
4. $\dfrac{1130}{256} = 4.4$ feet   5. $\dfrac{1130}{2} = 565$ Hz
6. four
7. $I = \dfrac{1}{D^2} = \dfrac{1}{100}$
   a. $\dfrac{1}{4} \times \dfrac{1}{100} = \dfrac{1}{400}$ $D = \sqrt{400} = 20$
   b. 40   c. 50
8. intensity

6-7 1. reflected, interface, walls, floors, ceilings
2. barrier   5. reverberation
3. equal to   6. secondary
4. a. source   7. diffraction
   b. spherical
   c. wall, image of the source
8. shadows, very little, because sound waves diffract to fill up the shadow, you don't need to be in the direct line of a sound.
9. diffract, sound source

6-8 1. superposition   2. interference

6-9 1. ray of sound   17. standing wave, stationary wave
2. wave front
3. inverse square law   18. wavelength
4. complex sound   19. frequency
5. fundamental frequency   20. elasticity
6. partial   21. elastic medium
7. overtone   22. amplitude spectrum
8. average power   23. beats
9. harmonic   24. pure tone
10. steady-state sound   25. intensity
11. masking   26. octave
12. transient sound   27. noise
13. noise   28. noise
14. white noise   29. waveform
15. incident wave   30. law of reflection
16. reflected wave

6-10 1. turbulence   2. aperiodically

6-11 1. filter   2. tuning   3. increased

6-12 1. power

6-13 1. interval   15. specifies smaller intensity ratios
2. ratio
3. 3   16. power, intensity
4. 10   17. area, time
5. 3:1   18. power
6. 10:1   19. power
7. (a) and (b)   20. 10^{-4}
8. logarithmic   21. pressure
9. interval   22. pressure
10. 5, exponents   23. area
11. 1, exponents   24. increase
12. 14   26. sound pressure
13. 10^{14}:1, 14   27. sound pressure level, SPL
14. tenth

6-14 1. 15–16 Hz to about 20,000 Hz
2. external, middle, inner
3. outer, inner; outer–protection; absorption and transformation of acoustic wave energy into mechanical vibratory energy; inner–absorption and transformation of mechanical energy into a series of neural impulses

6-15 1. conduct sound to tympanic membrane
2. isthmus   3. bony
4. lateral
5. in adults axis of canal is lower at orifice than it is medially; in children canal is more horizontal and also straighter
6. cerumen, keeps ear canal from drying out, prevents intrusion of insects and foreign bodies, ceruminous (modified sebaceous)
7. laterally   10. localization
8. azimuth   11. length
9. reflected   12. 4000
13. because compliance of the drum membrane modifies the effective length of the tube and generates a damping effect
14. higher   15. 15

6-16 1. a. right
   b. 8000 Hz because the head would be more of an obstacle to shorter sound waves and they would cast larger "sound shadows"

6-17 1. external auditory meatus, tympanic cavity
2. B, D, The tympanic membrane achieves full growth prenatally, but the meatus does not. As the meatus grows the angle will gradually change.
3. cone-shape, compliance of membrane

6-17 4. periphery, bony   8. umbo
cont'd 5. superiorly, limp   9. malleus
         6. concave   10. malleolar stria,
         7. malleolar stria   umbo

6-18 1. petrous, temporal   2. attic, epitympanic
         3. proper
         4. posterior, mastoid air, because of
            continuous mucous membrane lining
         5. meninges   8. chorda
         6. jugular   9. auditory
         7. bony   (Eustachian)

6-19 1. tympanic cavity, nasopharynx
         2. 1.3
         3. downward, forward, in children medial-
            ward, because it is more horizontal in
            children infections more easily transmitted
            from pharynx to middle ear
         4. equalize pressure and permit drainage to
            nasopharynx
         5. exudate, opening of the tube is above the
            floor of the cavity
         6. adenoids (pharyngeal tonsils)
         7. mucous membrane, negative
         8. zero, at + or − pressures movement of
            tympanic membrane and thus the ossicles is
            restricted; a damping effect.

6-20 1. prenatally
         2. efficient delivery of vibrations to inner ear
            fluids; protect inner ear from being
            overdriven by very strong vibrations
         3. annular   5. incudostapedial
         4. malleus, incus, even 6. distortion
         7. True, suspension system results in very little
            inertia and the ossicles are well balanced; if
            not ossicular chain would swing like a
            pendulum long after cessation of sound
         8. covered with mucous membrane
            continuous with lining of cavity

6-21 1. opposite   4. latency
         2. perpendicular   5. impedance
         3. reflex, acoustic reflex
         6. No, it protects us from most of the sounds of
            nature, but it cannot protect us from
            manmade sounds
         7. antagonists, synergists 8. intensity
         9. muscle or reflexive contraction
        10. a. False   10. d. False
            b. True   e. True
            c. False   f. False
        11. b   12. stapes

6-22 1. usually most of the sound will be reflected
            away
         2. elasticity and density
         3. impedance
         4. tympanic a&d, oval b&c
         5. airborne conduction across the middle ear
            space, bone conduction through the skull
            to the cochlea and inner ear fluids, trans-
            mission of mechanical vibration through
            the ossicular chain to the stapes footplate.
         6. a. theoretically 30, but may actually be 60
            b. pressure is equalized on both sides of
               the membrane
            c. 30 dB
         7. low, lower   11. pressure
         8. high   12. force
         9. c   13. same
        10. its periphery is fixed 14. different
            and the membrane   15. 2500
            is cone-shaped

6-23 1. equilibrium, hearing
         2. membranous, bony (osseous)
         3. blood, extracellular, intracellular
         4. both   5. vestibule

6-24 1. movement of the head and body
         2. static   3. kinetic

6-25 1. (1) scala vestibuli, (4) scala tympani
         2. (2) scala media
         3. (1) scala vestibuli
         4. (1) scala vestibuli, (4) scala tympani
         5. (5) basilar membrane
         6. (5) basilar membrane, (6) spiral lamina
         7. (7) vestibular membrane
         8. (6) spiral lamina
         9. (3) spiral ligament
        10. (6) spiral lamina
        11. (5) basilar membrane

6-26 *Arrangement of cells:*
         1. border cells of Held
         2. inner phalangeal cells
         3. inner rods (pillars) of Corti
         4. outer rods (pillars) of Corti
         5. outer phalangeal (Deiters') cells
         6. cells of Henson
         7. cells of Claudius

         1. outer phalangeal (Deiters')
         2. inner and outer rods of Corti
         3. reticular   6. outer, inner
         4. cells of Claudius   7. perforata
         5. border cells of Held

6-26   8. phalangeal,
cont'd   phalangeal
       9. endolymph, nerve
     10. reticular, tectorial
     14. mechanical
     15. cover net, fibrous main body, homogeneous basal layer (of Hardesty)
     16. prenatally, when hearing begins contact is probably forcefully weakened
     17. Henson's

     11. media, endolymph
     12. supportive, receptive
     13. stereocilia

6-27   1. neural
     2. closed, unyielding, are not, fluids
     3. fluid column of the cochlea responds as a whole (mass-action mechanism), pressure generated in the scala vestibuli is transmitted across scala media to scala tympani
     4. place b&d, frequency a&c
     5. pitch or frequency
     6. basilar membrane fibers are not resonating elements
     7. radial, unique
     8. basal
     9. analytic
     10. transducer
     11. transducer, neural
     12. brain
     13. telephone
     14. that they were capable of discharging an extraordinary number of impulses per second
     15. none
     16. refute, causes frequency dependent localized hair cell damage, thus supporting an analytic function of the inner ear
     17. membrane not under appreciable longitudinal tension, thus deformation of a segment will not affect the entire membrane
     18. when using the correct values Bekesy was unable to generate the standing waves observed by Ewald; either standing waves or traveling waves may be generated on the membrane
     19. frequency, basal, apical
     20. stapes footplate
     21. traveling wave, not a true traveling wave (as on a string); may not reflect the actual displacement pattern on the basilar membrane in response to a progressive pressure gradient within the cochlear fluids
     22. instantaneously   23. gelatin, interface
     24. interface, time, space, scala, vibrates
     25. without this release of pressure, stapes could not vibrate

6-28   1. physical
     2. 0.1, 0.5
     3. apex, base
     4. yes
     5. C B A

6-28   6. a. frequency
cont'd   b. zero, membrane
       c. they displace much larger segments of the membrane
       d. increases
       e. will not
       f. complex
       g. amplitude
       h. place
       i. stiffness
       j. because of the exchange of energy between the membrane and the cochlear fluids
       k. see no. 1

6-29   1. 4,000; 15,000
     2. no
     3. volley
     4. place, frequency
     5. number of active fibers and the rates at which they act
     6. yes

6-30   1. compressed, stretched, shearing
     2. different, force of shear
     3. tectorial
     4. endolymph
     5. displacement, velocity
     6. electrical, hair cell body, support

6-31   1. transducer
     2. stereocilia, endolymph
     3. a. True   3. c. False
       b. True   d. True
     4. a. media, tympani, acoustic
       b. separated
       c. outer, microphonics
     5. a. potential, microphone
       b. hair cells   c. high
       d. distortion free, no real threshold, no adaptation to stimulus, no fatigue, no frequency limits (within reason)
       e. help determine functional integrity of inner ear and the location of frequency-dependent receptors on the basilar membrane
       f. tonotopographical
       g. study hair cells damaged by exposure to intense sound
       h. (1) apical   (4) volley
           (2) basal   (5) place
           (3) low
     6. a. cochlea, acoustic (auditory)
       b. no   c. impulses
       d. a click or a high frequency tone burst
       e. raised

6-32   1. False
     2. False
     3. True
     4. True
     5. True

6-33
1. frequency
2. bioelectrical
   a. habenula
   b. myelinated
   c. node
   d. cerebral cortex
3. neural
4. a. increased
   b. increased
5. a. degree
   b. cochlear microphonics
   c. summating potentials
6. resting
7. generator, nerve
8. a. tuning b. wide, best c. True

9. a. linear
   b. nonlinear
   c. nonlinear
   d. linear
   e. decreases
10. all True (a–i)
11. a. spontaneous, threshold
    b. increases
    c. best or characteristic, low
    d. phase-locked
    e. nerve fiber, repeated
    f. interval
    g. phase-locked

6-34
1. acoustic (auditory), vestibular, cochlear
2. thick, small   3. homeostasis

6-35
1. cerebrum   2. cochlear
3. apical, basal, nerve developed before cochlea begins its spiral growth, "dragging along" basal nerve fibers

6-35
cont'd
4. higher
5. vestibular
6. cochlear, in region of medulla oblongata
7. both, destruction of one auditory pathway does not cause complete deafness
8. lemniscus   11. cerebral cortex, cochlea
9. reflexive
10. cerebral cortex   12. integrative

6-36
1. when there is direct contact between skull and vibrating body; when airborne sound is extremely intense
2. some of the low frequencies you hear via bone conduction never become airborne
3. cancellation, the same
4. labyrinthine bone conduction, inertial lag of ossicular chain, occlusion effect
5. the same   8. True
6. vestibuli, tympani   9. augment
7. labyrinth, endolymph, vestibuli, heightened
10. frequency, phase; results in displacement of cartilaginous skeleton of canal thus generating airborne sounds which follow conventional route

Chapter 7

7-1
1. a. myotome   c. dermatome
   b. sclerotome
2. The level of origin of a muscle is reflected by the segmental level of its nerve supply. The nerve supply to a developing muscle is acquired early from nerves arising at the same segmental level. When muscle migration occurs the already attached nerve is simply pulled along.
3. parts of, or entire muscle segments which may degenerate during embryological development tend to convert to connective tissue, thus forming aponeuroses.

7-2
1. about end of 3rd week
2. prosencephaton
3. face   4. oral groove
5. branchial, branchial, gill
6. nasal
7. frontonasal and maxillary processes, mandibular and hyoid arches
8. oronasal   10. philtrum, globular
9. nasooptic   11. lateral

7-3
1. oral, nasal, maxillary
2. choanae, oral   5. palate
3. anterior   6. tongue
4. mouth, vertical
7. the tongue moves downward, evacuating the space between the palatine processes
8. horizontal, anteriorly
9. septum   11. medial
10. lateral   12. premaxillary portion

7-5
1. ectoderm   5. nerve, supportive
2. mesoderm, vertebral column, skull   6. inner to outer neural crest
3. ventricular, central canal   7. myelin
4. peripheral   8. spinal cord
9. The number of dendrites in the marginal layer increase leaving less room
10. 2
11. afferent (sensory), efferent (motor)
12. internuncial

Answers to Selected Questions

No. 1-3, page 4

belly button

No. 2-6, page 35

After a person has smoked one cigarette the cilia will be non-motile for several hours.

No. 2-37, page 70

Residual air is the air remaining in the lungs after you have exhaled as much air as possible. You cannot speak on air that cannot be exhaled.

No. 2-41, page 75

Inadequate velopharyngeal closure will prevent the build-up of intraoral pressure, but does not affect the build-up of pressure at the laryngeal level. Because the vocal folds then offer the most effective means of achieving resistance to air flow, glottal stops and attacks are sometimes substituted for stops and fricatives.

Enlarged adenoids, engorged mucous membrane, polyps, and severe deviation of the nasal septum are examples of causes of increased air flow resistance in the nasal passageways. Hyponasality and/or mouthbreathing may be the result.

No. 3-7, page 83

Although the measured thyroid angle of males and females is about the same, in outward appearance the male larynx is quite angular and the female larynx is more rounded. Also, the male larynx grows very rapidly during puberty.

No. 3-24, page 106

You pucker your lips to restrict the airstream, thereby increasing its velocity.

No. 3-25, page 107

Glottal attacks are used phonemically in some language dialects. The way in which the attack is initiated probably determines whether or not it is abusive.

Speech production does not occur in isolation, but is affected by many aspects of behavior and thus often reflects some primary personality characteristics.

High-speed cinematography is rarely used in diagnosis because

1. it requires training of the subject (often many weeks).
2. it is extremely expensive.
3. it covers only about one second of vibratory movement.
4. it does not reveal information about the dynamics of vocal fold vibration.
5. fiberoptic endoscopy permits observation during connected speech.

No. 4-7, page 126

One of the first signs of aging is graying at the temples. Could it be that the gray matter from the brain is escaping into the hair?

No. 4-9, page 134

Infection or inflamation of the sinuses and air cells may spread to the meninges.

No. 4-27, page 161

The presence or absence of Passavant's pad cannot be determined during an oral examination because, if present, it is located above the level of the soft palate.

No 4-34, page 170

The vocal tract of an adult male and the larynx of a child would produce a high-pitch with adult formant frequencies, and the voice would resemble that of a munchkin in the *Wizard of Oz*.

No. 5-6, page 184

Meningitis is an inflamation of the meninges.

The brain floats in cerebrospinal fluid.

It will result in an extradural hematoma (a localized mass of blood, usually partially clotted, found outside a blood vessel). Continued bleeding may lead to coma or death.

Fig. 5.4, page 187

Flotation of the brain in cerebrospinal fluid reduces its effective weight.

No. 6-36, page 277

At frequencies above 2,000 Hz the vibrations are too small to have the same effect as those below 2,000 Hz.